1,000,000 Books

are available to read at

Forgotten Books

www.ForgottenBooks.com

Read online
Download PDF
Purchase in print

ISBN 978-1-332-32656-3
PIBN 10314471

This book is a reproduction of an important historical work. Forgotten Books uses state-of-the-art technology to digitally reconstruct the work, preserving the original format whilst repairing imperfections present in the aged copy. In rare cases, an imperfection in the original, such as a blemish or missing page, may be replicated in our edition. We do, however, repair the vast majority of imperfections successfully; any imperfections that remain are intentionally left to preserve the state of such historical works.

Forgotten Books is a registered trademark of FB &c Ltd.
Copyright © 2018 FB &c Ltd.
FB &c Ltd, Dalton House, 60 Windsor Avenue, London, SW19 2RR.
Company number 08720141. Registered in England and Wales.

For support please visit www.forgottenbooks.com

1 MONTH OF FREE READING

at

www.ForgottenBooks.com

By purchasing this book you are eligible for one month membership to ForgottenBooks.com, giving you unlimited access to our entire collection of over 1,000,000 titles via our web site and mobile apps.

To claim your free month visit:
www.forgottenbooks.com/free314471

* Offer is valid for 45 days from date of purchase. Terms and conditions apply.

English
Français
Deutsche
Italiano
Español
Português

www.forgottenbooks.com

Mythology Photography **Fiction** Fishing Christianity **Art** Cooking Essays Buddhism Freemasonry Medicine **Biology** Music **Ancient Egypt** Evolution Carpentry Physics Dance Geology **Mathematics** Fitness Shakespeare **Folklore** Yoga Marketing **Confidence** Immortality Biographies Poetry **Psychology** Witchcraft Electronics Chemistry History **Law** Accounting **Philosophy** Anthropology Alchemy Drama Quantum Mechanics Atheism Sexual Health **Ancient History Entrepreneurship** Languages Sport Paleontology Needlework Islam **Metaphysics** Investment Archaeology Parenting Statistics Criminology **Motivational**

FLYING MACHINES TODAY

BY

WILLIAM DUANE ENNIS

Professor of Mechanical Engineering in the Polytechnic Institute of Brooklyn

123 ILLUSTRATIONS

NEW YORK
D. VAN NOSTRAND COMPANY
23 Murray and   1911   27 Warren Sts.

Copyright, 1911, by
D. VAN NOSTRAND COMPANY

THE · PLIMPTON · PRESS · NORWOOD · MASS · U · S · A

To
MY MOTHER

223083

PREFACE

SPEAKING with some experience, the writer has found that instruction in the principles underlying the science and sport of aviation must be vitalized by some contemporaneous study of what is being accomplished in the air. No one of the revolutionizing inventions of man has progressed as rapidly as aerial navigation. The "truths" of today are the absurdities of tomorrow.

The suggestion that some grasp of the principles and a very fair knowledge of the current practices in aeronautics may be had without special technical knowledge came almost automatically. If this book is comprehensible to the lay reader, and if it conveys to him even a small proportion of the writer's conviction that flying machines are to profoundly influence our living in the next generation, it will have accomplished its author's purpose.

POLYTECHNIC INSTITUTE OF BROOKLYN,
    NEW YORK, April, 1911.

CONTENTS

	PAGE
THE DELIGHTS AND DANGERS OF FLYING. — Dangers of Aviation. — What it is Like to Fly	1
SOARING FLIGHT BY MAN. — What Holds it Up? — Lifting Power. — Why so Many Sails? — Steering	17
TURNING CORNERS. — What Happens when Making a Turn. — Lateral Stability. — Wing Warping. — Automatic Control. — The Gyroscope. — Wind Gusts.	33
AIR AND THE WIND. — Sailing Balloons. — Field and Speed	43
GAS AND BALLAST. — Buoyancy in Air. — Ascending and Descending. — The Ballonet. — The Equilibrator	57
DIRIGIBLE BALLOONS AND OTHER KINDS. — Shapes. — Dimensions. — Fabrics. — Framing. — Keeping the Keel Horizontal. — Stability. — Rudders and Planes. — Arrangement and Accessories. — Amateur Dirigibles. — The Fort Omaha Plant. — Balloon Progress	71
THE QUESTION OF POWER. — Resistance of Aeroplanes. — Resistance of Dirigibles. — Independent Speed and Time-Table. — The Cost of Speed. — The Propeller	101
GETTING UP AND DOWN; MODELS AND GLIDERS; AEROPLANE DETAILS. — Launching. — Descending. — Gliders. — Models. — Balancing. — Weights. — Miscellaneous. — Things to Look After	121
SOME AEROPLANES. — SOME ACCOMPLISHMENTS	143
THE POSSIBILITIES IN AVIATION. — The Case of the Dirigible. — The Orthopter. — The Helicopter. — Composite Types. — What is Promised	170
AERIAL WARFARE	189

LIST OF ILLUSTRATIONS

	PAGE
The Fall of Icarus	*Frontispiece*
The Aviator	3
The Santos-Dumont "*Demoiselle*"	4
View from a Balloon	9
Anatomy of a Bird's Wing	10
Flight of a Bird	11
In a Meteoric Shower	13
How a Boat Tacks	15
Octave Chanute	18
Pressure of the Wind	19
Forces Acting on a Kite	20
Sustaining Force in the Aeroplane	23
Direct Lifting and Resisting Forces	24
Shapes of Planes	26
Balancing Sail	28
Roe's Triplane at Wembley	30
Action of the Steering Rudder	31
Recent Type of Wright Biplane	31
Circular Flight	33
The Aileron	35
Wing Tipping	36
Wing Warping	37
The Gyroscope	39
Diurnal Temperatures at Different Heights	45
Seasonal Variation in Wind Velocities	47
The Wind Rose for Mt. Weather, Va.	49
Diagram of Parts of a Drifting Balloon	51
Glidden and Stevens Getting Away in the "*Boston*"	52
Relative and Absolute Balloon Velocities	53
Field and Speed	53
Influence of Wind on Possible Course	54
Count Zeppelin	55
Buoyant Power of Wood	57
One Cubic Foot of Wood Loaded in Water	58

List of Illustrations

	PAGE
Buoyant Power of Hydrogen	59
Lebaudy's "*Jaune*"	60
Air Balloon	62
Screw Propeller for Altitude Control	66
Balloon with Ballonets	67
Construction of the Zeppelin Balloon	68
The Equilibrator	69
Henry Giffard's Dirigible	71
Dirigible of Dupuy de Lome	72
Tissandier Brothers' Dirigible Balloon	73
The "*Baldwin*"	74
The "*Zeppelin*" on Lake Constance	75
The "*Patrie*"	77
Manufacturing the Envelope of a Balloon	79
Andrée's Balloon, "*L'Oernen*"	80
Wreck of the "*Zeppelin*"	82
Car of the "*Zeppelin*"	84
Stern View of the "*Zeppelin*"	86
The "*Clément-Bayard*"	87
The "*Ville de Paris*"	88
Car of the "*Liberté*"	89
The "*Zodiac No. 2*"	92
United States Signal Corps Balloon Plant at Fort Omaha	93
The "*Caroline*"	94
The Ascent at Versailles, 1783	95
Proposed Dirigible	96
The "*République*"	97
The First Flight for the Gordon-Bennett Cup	99
The Gnome Motor	102
Screw Propeller	103
One of the Motors of the "*Zeppelin*"	104
The Four-Cycle Engine	105
Action of Two-Cycle Engine	106
Motor and Propeller	108
Two-Cylinder Opposed Engine	110
Four-Cylinder Vertical Engine	110
Head End Shapes	113
The Santos-Dumont Dirigible No. 2	115
In the Bay of Monaco: Santos-Dumont	117
Wright Biplane on Starting Rail	121
Launching System for Wright Aeroplane	122

List of Illustrations

	PAGE
The Nieuport Monoplane	124
A Biplane	125
Ely at Los Angeles	126
Trajectory During Descent	127
Descending	128
The Witteman Glider	130
French Monoplane	132
A Problem in Steering	133
Lejeune Biplane	134
Tellier Monoplane	135
A Monoplane	137
Cars and Framework	139
Some Details	139
Recent French Machines	141
Orville Wright at Fort Myer	143
The First Flight Across the Channel	144
Wright Motor	145
Voisin-Farman Biplane	147
The Champagne Grand Prize Flight	148
Farman's First Biplane	149
The "*June Bug*"	150
Curtiss Biplane	151
Curtiss' Hydro-Aeroplane at San Diego Bay	152
Flying Over the Water	153
Bleriot-Voisin Cellular Biplane with Pontoons	154
Latham's "*Antoinette*"	155
James J. Ward at Lewiston Fair	156
Marcel Penot in the "*Mohawk*"	157
Santos-Dumont's "*Demoiselle*"	159
Blériot Monoplane	160
Latham's Fall into the Channel	161
De Lesseps Crossing the Channel	163
The Maxim Aeroplane	164
Langley's Aeroplane	165
Robart Monoplane	166
Vina Monoplane	167
Blanc Monoplane	170
Melvin Vaniman Triplane	171
Jean de Crawhez Triplane	171
A Triplane	172
Giraudon's Wheel Aeroplane	175

List of Illustrations

	PAGE
Bréguet Gyroplane (Helicopter)	177
Wellman's "*America*"	181
The German Emperor Watching the Progress of Aviation	189
Automatic Gun for Attacking Airships	193
Gun for Shooting at Aeroplanes	197
Santos-Dumont Circling the Eiffel Tower	199
Latham, Farman and Paulhan	202

FLYING MACHINES TODAY

THE DELIGHTS AND DANGERS OF FLYING

FEW things have more charm for man than flight. The soaring of a bird is beautiful and the gliding of a yacht before the wind has something of the same beauty. The child's swing; the exercise of skating on good ice; a sixty-mile-an-hour spurt on a smooth road in a motor car; even the slightly passé bicycle: these things have all in their time appealed to us because they produce the illusion of flight — of progress through the intangible air with all but separation from the prosaic earth.

But these sensations have been only illusions. To actually leave the earth and wander at will in aerial space — this has been, scarcely a hope, perhaps rarely even a distinct dream. From the days of Dædalus and Icarus, of Oriental flying horses and magic carpets, down to "Darius Green and his flying machine," free flight and frenzy were not far apart. We were learnedly told, only a few years since, that sustention by heavier-than-air machines was impossible without the discovery, first, of some new matter or some new force. It is now (1911) only eight years since Wilbur Wright at Kitty Hawk, with the aid of the new (?) matter — aluminum — and the "new" force — the gasoline engine — in three successive flights proved that a man could travel through the air and safely descend, in a machine weighing many times as much as the air it displaced. It is only five years since two designers — Surcouf and Lebaudy — built dirigible balloons approximating present forms, the *Ville de Paris* and *La Patrie*. It is only now that we average people may confidently contemplate the prospect of an aerial voyage for ourselves before we die. A contemplation not without its shudder, perhaps; but yet not altogether more daring than that of our grandsires who first rode on steel rails behind a steam locomotive.

The Dangers of Aviation

We are very sure to be informed of the fact when an aviator is killed. Comparatively little stir is made nowadays over an automobile fatality, and the ordinary railroad accident receives bare mention. For instruction and warning, accidents to air craft cannot be given too much publicity; but if we wish any accurate conception of the danger we must pay regard to factors of proportion. There are perhaps a thousand aeroplanes and about sixty dirigible balloons in the world. About 500 men — amateurs and professionals — are continuously engaged in aviation. The Aero Club of France has issued in that country nearly 300 licenses. In the United States, licenses are held by about thirty individuals. We can form no intelligent estimate as to the number of unlicensed amateurs of all ages who are constantly experimenting with gliders at more or less peril to life and limb.

A French authority has ascertained the death rate

The Delights and Dangers of Flying

among air-men to have been — to date — about 6%. This is equivalent to about one life for 4000 miles of flight: but we must remember that accidents will vary rather with the number of ascents and descents than with the mileage. Four thousand miles in 100 flights would be

much less perilous, under present conditions, than 4000 miles in 1000 flights.

There were 26 fatal aeroplane accidents between September 17, 1908, and December 3, 1910. Yet in that period there were many thousands of ascents: 1300 were made in one week at the Rheims tournament alone. Of the 26 accidents, 1 was due to a wind squall, 3 to collision, 6 (apparently) to confusion of the aviator, and 12 to mechanical breakage. An analysis of 40 British accidents shows 13 to have been due to engine failures, 10 to alighting on bad ground, 6 to wind gusts, 5 to breakage of the propeller, and 6 to fire and miscellaneous causes. These

THE SANTOS-DUMONT "DEMOISELLE"
(From *The Aeroplane*, by Hubbard, Ledeboer and Turner)

casualties were not all fatal, although the percentage of fatalities in aeronautic accidents is high. The most serious results were those due to alighting on bad ground; long grass and standing grain being very likely to trip the machine and throw the occupant. French aviators are now strapping themselves to their seats in order to avoid this last danger.

Practically all of the accidents occur to those who are flying; but spectators may endanger themselves. During one of the flights of Mauvais at Madrid, in March of the present year, the bystanders rushed through the barriers and out on the field before the machine had well started. A woman was decapitated by the propeller, and four other persons were seriously injured.

Nearly all accidents result from one of three causes: bad design, inferior mechanical construction, and the taking of unnecessary risks by the operator. Scientific design at the present writing is perhaps impossible. Our knowledge of the laws of air resistance and sustention is neither accurate nor complete. Much additional study and experiment must be carried on; and some better method of experimenting must be devised than that which sends a man up in the air and waits to see what happens. A thorough scientific analysis will not only make aviation safer, it will aid toward making it commercially important. Further data on propeller proportions and efficiencies, and on strains in the material of screws under aerial conditions will do much to standardize power plant equipment. The excessive number of engine breakdowns is obviously related to the extremely light weight of the engines employed: better design may actually increase these weights over those customary at present. Great weight reduction is no longer regarded as essential at present speeds in aerial navigation: we have perhaps already gone too far in this respect.

Bad workmanship has been more or less unavoidable, since no one has yet had ten years' experience in building aeroplanes. The men who have developed the art have usually been sportsmen rather than mechanics, and only time is necessary to show the impropriety of using "safety pins" and bent wire nails for connections.

The taking of risks has been an essential feature. When one man earns $100,000 in a year by dare-devil flights, when the public flocks in hordes — and pays good prices — to see a man risk his neck, he will usually aim to satisfy it. This is not developing aerial navigation: this is circus riding — looping-the-loop performances which appeal to some savage instinct in us but lead us nowhere. Men have climbed two miles into the clouds, for no good purpose whatever. All that we need to know of high altitude conditions is already known or may be learned by ascents in anchored balloons. Records up to heights of sixteen miles have been obtained by sounding balloons.

If these high altitudes may under certain conditions be desirable for particular types of balloon, they are essentially undesirable for the aeroplane. The supporting power of a heavier-than-air machine decreases in precisely inverse ratio with the altitude. To fly high will then involve either more supporting surface and therefore a structurally weaker machine, or greater speed and consequently a larger motor. It is true that the resistance to propulsion decreases at high altitudes, just as the supporting power decreases: and on this account, given only a sufficient margin of supporting power, we might expect a standard machine to work about as well at a two-mile elevation as at a height of 200 feet; but rarefaction of the air at the higher altitudes decreases the weight of carbureted mixture drawn into the motor, and consequently its output. Any air-man who attempts to reach great heights in a machine not built for such purpose is courting disaster.

Flights over cities, spectacular as they are, and popular as they are likely to remain, are doubly dangerous on account of the irregular air currents and absence of safe landing places. They have at last been officially discountenanced as not likely to advance the sport.

All flights are exhibition flights. The day of a quiet, mind-your-own-business type of aerial journey has not yet arrived. Exhibition performances of any sort are generally hazardous. There were nine men killed in one recent automobile meet. If the automobile were used exclusively for races and contests, the percentage of fatalities might easily exceed that in aviation. It is claimed that no inexperienced aviator has ever been killed. This may not be true, but there is no doubt that the larger number of accidents has occurred to the better-known men from whom the public expects something daring.

Probably the best summing up of the danger of aviation may be obtained from the insurance companies. The courts have decided that an individual does not forfeit his life insurance by making an occasional balloon trip. Regular classified rates for aeroplane and balloon operators are in force in France and Germany. It is reported that Mr. Grahame-White carries a life insurance policy at 35% premium — about the same rate as that paid by a "crowned head." Another aviator of a less professional type has been refused insurance even at 40% premium. Policies of insurance may be obtained covering damage to machines by fire or during transportation and by collisions with other machines; and covering liability for injuries to persons other than the aviator.

On the whole, flying is an ultra-hazardous *occupation;* but an *occasional* flight by a competent person or by a passenger with a careful pilot is simply a thrilling experience, practically no more dangerous than many things we do without hesitation. Nearly all accidents have been due to preventable causes; and it is simply a matter of science, skill, perseverance, and determination to make an aerial excursion under proper conditions as safe as a journey in a motor car. Men who for valuable prizes undertake spectacular feats will be killed as frequently in aviation as in bicycle or even in automobile racing; but probably not very much more frequently, after design and workmanship in flying machines shall have been perfected. The total number of deaths in aviation up to February 9, 1911, is stated to have been forty-two.

What It Is Like to Fly

We are fond of comparing flying machines with birds, with fish, and with ships: and there are useful analogies

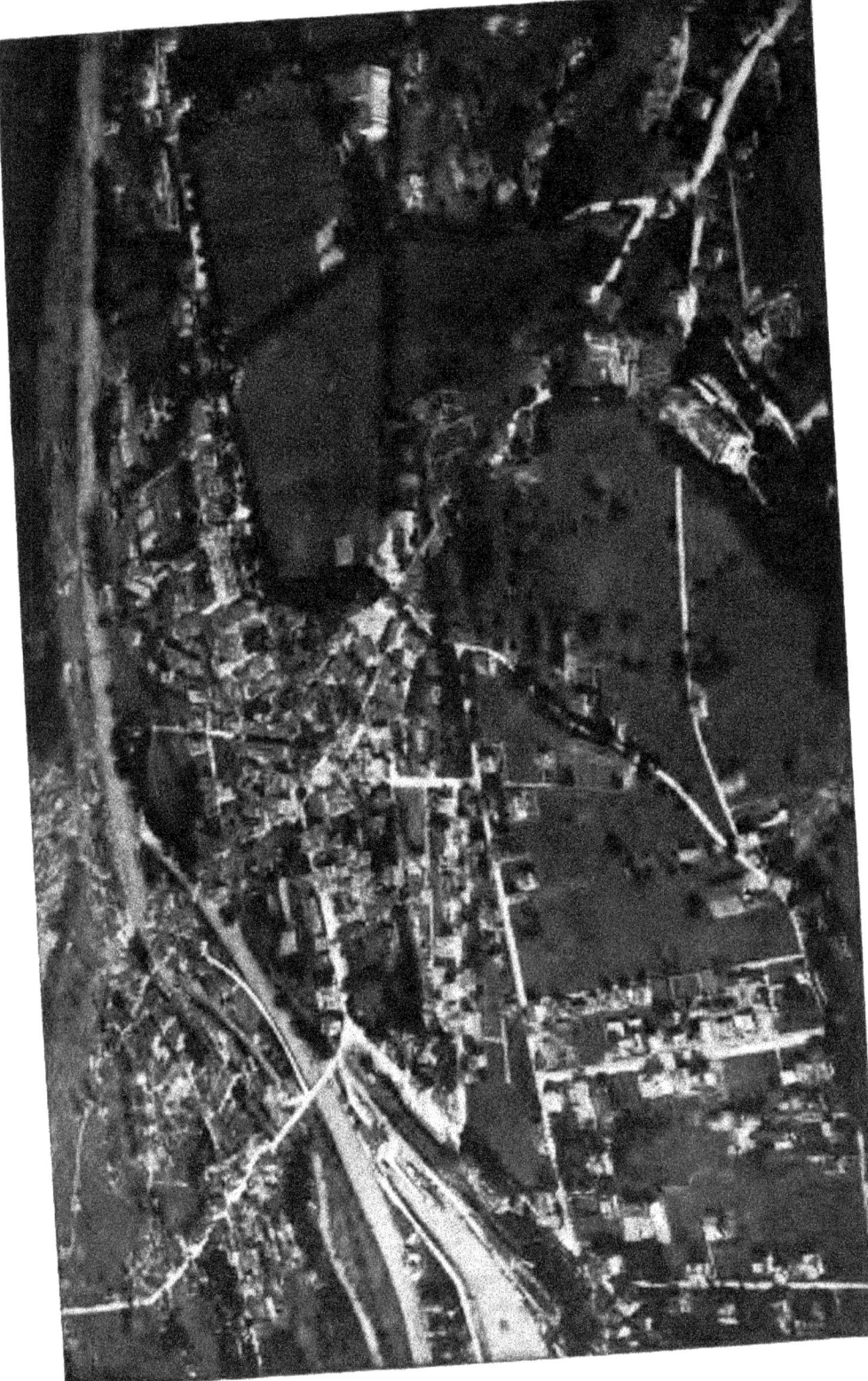

with all three. A drifting balloon is like a becalmed ship or a dead fish. It moves at the speed of the aerial fluid about it and the occupants perceive no movement whatever. The earth's surface below appears to move in the opposite direction to that in which the wind carries the balloon. With a dirigible balloon or flying machine, the sensation is that of being exposed to a violent wind, against which (by observation of landmarks) we find that we

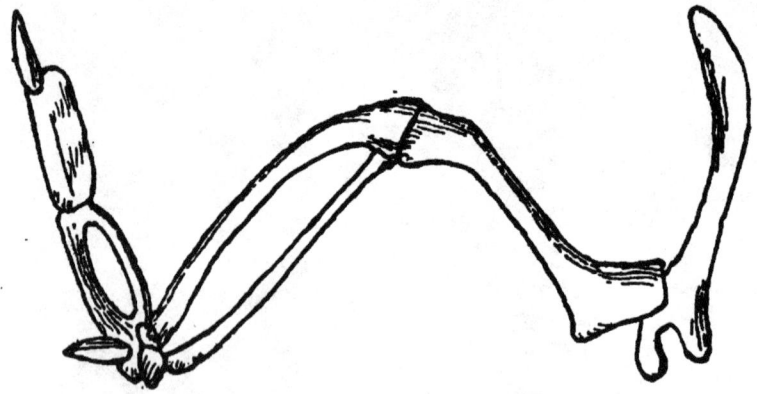

ANATOMY OF A BIRD'S WING
(From Walker's *Aerial Navigation*)

progress. It is the same experience as that obtained when standing in an exposed position on a steamship, and we wonder if a bird or a fish gradually gets so accustomed to the opposing current as to be unconscious of it. But in spite of jar of motors and machinery, there is a freedom of movement, a detachment from earth-associations, in air flight, that distinguishes it absolutely from the churning of a powerful vessel through the waves.

Birds fly in one of three ways. The most familiar bird

flight is by a rapid wing movement which has been called oar-like, but which is precisely equivalent to the usual movement of the arms of a man in swimming. The edge of the wing moves forward, cutting the air; on the return stroke the leading edge is depressed so as to present a nearly flat surface to the air and thus propel the bird forward. A slight downward direction of this stroke serves to impel the flight sufficiently upward to offset the effect of gravity. Any man can learn to swim, but no man can fly, because neither in his muscular frame nor by any device which he can attach thereto can he exert a sufficient pressure to overcome his own weight against as imponderable a fluid as air. If air were as heavy as water, instead of 700 times lighter, it would be as easy to fly as to swim. The bird can fly because of the great surface, powerful construction, and rapid movement of its wings, in proportion to the weight of its body. But compared with the rest of the animal kingdom, flying birds are all of small size. Helmholz considered that the vulture represented the heaviest body that could possibly be raised and kept aloft by the exercise of muscular power, and it is understood that vultures have considerable difficulty in ascending; so much so that unless in a position to take a short preliminary run they are easily captured.

Every one has noticed a second type of bird flight — soaring. It is this flight which is exactly imitated in a glider. An aeroplane differs from a soaring bird only in that it carries with it a producer of forward impetus — the propeller — so that the soaring flight may last indefinitely: whereas a soaring bird gradually loses speed and descends.

IN A METEORIC SHOWER

A third and rare type of bird flight has been called *sailing*. The bird faces the wind, and with wings outspread and their forward edge elevated rises while being forced backward under the action of the breeze. As soon as the wind somewhat subsides, the bird turns and *soars* in the desired direction. Flight is thus accomplished without muscular effort other than that necessary to properly incline the wings and to make the turns. It is practicable only in squally winds, and the birds which practice "sailing" — the albatross and frigate bird — are those which live in the lower and more disturbed regions of the atmosphere. This form of flight has been approximately imitated in the manœuvering of aeroplanes.

Comparison of flying machines and ships suggests many points of difference. Water is a fluid of great density, with a definite upper surface, on which marine structures naturally rest. A vessel in the air may be at any elevation in the surrounding rarefied fluid, and great attention is necessary to keep it at the elevation desired. The air has no surface. The air ship is like a submarine — the dirigible balloon of the sea — and perhaps rather more safe. An ordinary ship is only partially immersed; the resistance of the fluid medium is exerted over a portion only of its head end: but the submarine or the flying machine is wholly exposed to this resistance. The submarine is subjected to ocean currents of a very few miles per hour, at most; the currents to which the flying machine may be exposed exceed a mile a minute. Put a submarine in the Whirlpool Rapids at Niagara and you will have possible air ship conditions.

A marine vessel may *tack*, *i.e.*, may sail partially against the wind that propels it, by skilful utilization of the resist-

The Delights and Dangers of Flying   15

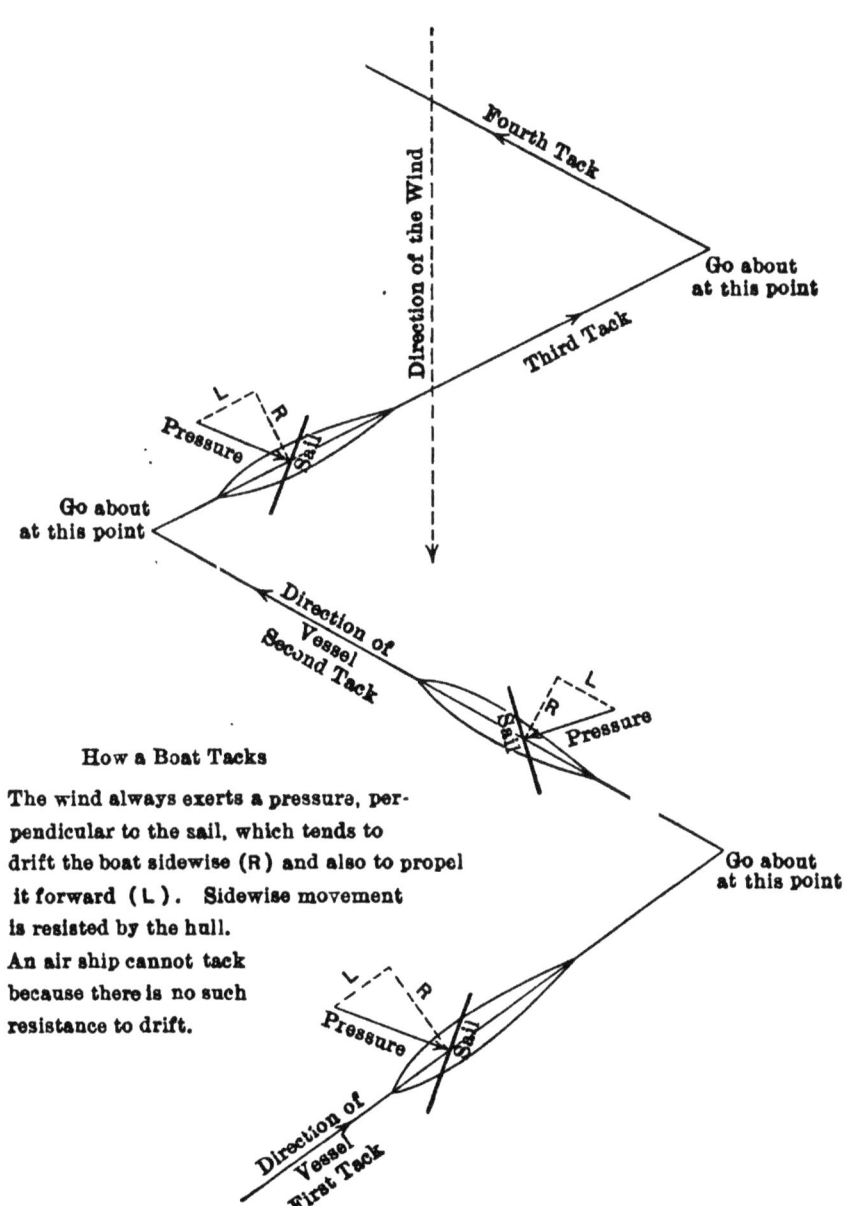

How a Boat Tacks

The wind always exerts a pressure, perpendicular to the sail, which tends to drift the boat sidewise (R) and also to propel it forward (L). Sidewise movement is resisted by the hull. An air ship cannot tack because there is no such resistance to drift.

ance to sidewise movement of the ship through the water: but the flying machine is wholly immersed in a single fluid, and a head wind is nothing else than a head wind, producing an absolute subtraction from the proper speed of the vessel.

Aerial navigation is thus a new art, particularly when heavier-than-air machines are used. We have no heavier-than-water *ships*. The flying machine must work out its own salvation.

SOARING FLIGHT BY MAN

FLYING machines have been classified as follows: —

> LIGHTER THAN AIR
> Fixed balloon,
> Drifting balloon,
> Sailing balloon,
> Dirigible balloon
> > rigid (Zeppelin),
> > ballonetted.
>
> HEAVIER THAN AIR
> Orthopter,
> Helicopter,
> Aeroplane
> > monoplane,
> > multiplane.

We will fall in with the present current of popular interest and consider the aeroplane — that mechanical grasshopper — first.

WHAT HOLDS IT UP?

When a flat surface like the side of a house is exposed to the breeze, the velocity of the wind exerts a force or pressure directly against the surface. This principle is taken into account in the design of buildings, bridges, and other

OCTAVE CHANUTE (died 1910)
To the researches of Chanute and Langley must be ascribed much of American progress in aviation.

structures. The pressure exerted per square foot of surface is equal (approximately) to the square of the wind velocity in miles per hour, divided by 300. Thus, if the wind velocity is thirty miles, the pressure against a house wall on which it acts directly is 30 × 30 ÷ 300 = 3 pounds per square foot: if the wind velocity is sixty miles, the pressure is 60 × 60 ÷ 300 = 12 pounds: if the velocity is ninety miles, the pressure is 90 × 90 ÷ 300 = 27 pounds, and so on.

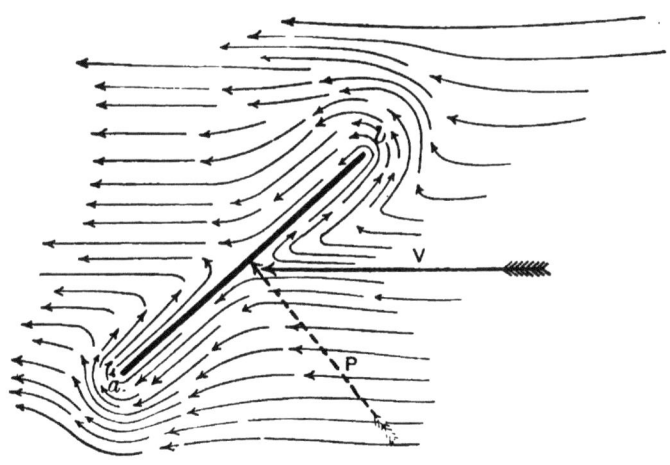

If the wind blows obliquely toward the surface, instead of directly, the pressure at any given velocity is reduced, but may still be considerable. Thus, in the sketch, let *ab* represent a wall, toward which we are looking downward, and let the arrow *V* represent the direction of the wind. The air particles will follow some such paths as those indicated, being deflected so as to finally escape around the ends of the wall. The result is that a pressure is produced which may be considered to act along the dotted line *P*, perpendicular to the wall. This is the invariable law: that no matter how oblique the surface may be, with reference to the direction of the wind, there is always a pressure produced against the surface by the wind, and this pressure always acts *in a direction perpendicular to the surface*. The amount of pressure will depend upon the wind velocity and the obliquity or inclination of the surface (*ab*) with the wind (*V*).

Now let us consider a kite — the "immediate ancestor" of the aeroplane. The surface *ab* is that of the kite itself,

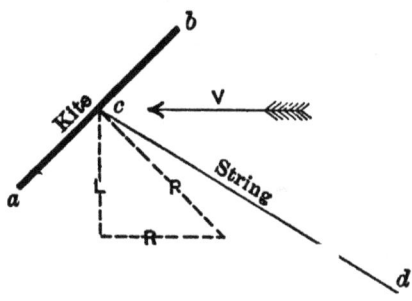

held by its string *cd*. We are standing at one side and looking at the *edge* of the kite. The wind is moving horizontally against the face of the kite, and produces a pressure *P* directly against the latter. The pressure tends both to move it toward the left and to lift it. If the tendency to move toward the left be overcome by the string, then the tendency toward lifting may be offset — and in practice *is* offset — by the weight of the kite and tail.

We may represent the two tendencies to movement produced by the force *P*, by drawing additional dotted lines, one horizontally to the left (*R*) and the other vertically (L); and it is known that if we let the length of the line P represent to some convenient scale the amount of direct pressure, then the lengths of R and L will also represent to the same scale the amounts of horizontal and vertical force due to the pressure. If the weight of kite and tail exceeds the vertical force L, the kite will descend: if these weights are less than that force, the kite will ascend. If they are precisely equal to it, the kite will neither ascend nor descend. The ratio of L to R is determined by the slope of P; and this is fixed by the slope of ab; so that we have the most important conclusion: *not only does the amount of direct pressure (P) depend upon the obliquity of the surface with the breeze (as has already been shown), but the relation of vertical force (which sustains the kite) to horizontal force also depends on the same obliquity.* For example, if the kite were flying almost directly above the boy who held the string, so that ab became almost horizontal, P would be nearly vertical and L would be much greater than R. On the other hand, if ab were nearly vertical, the kite flying at low elevation, the string and the direct pressure would be nearly horizontal and L would be much less than R. The force L which lifts the kite seems to increase while R decreases, as the kite ascends: but L may not actually increase, because it depends upon the amount of direct pressure, P, as well as upon the direction of this pressure; and the amount of direct pressure steadily decreases during ascent, on account of the increasing obliquity of ab with V. All of this is of course dependent on the assumption that the kite always has the same inclination to the string, and the described resolution of the forces, although answering for illustrative purposes, is technically incorrect.

It seems to be the wind velocity, then, which holds up the kite: but in reality the string is just as necessary as the wind. If there is no string, and the wind blows the kite with it, the kite comes down, because the pressure is wholly due to a relative velocity as between kite and wind. The wind exerts a pressure against the rear of a railway train, if it happens to be blowing in that direction, and if we stood on the rear platform of a stationary train we should feel that pressure: but if the train is started up and caused to move at the same speed as the wind there would be no pressure whatever.

One of the very first heavier-than-air flights ever recorded is said to have been made by a Japanese who dropped bombs from an immense man-carrying kite during the Satsuma rebellion of 1869. The kite as a flying machine has, however, two drawbacks: it needs the wind — it cannot fly in a calm — and it stands still. One early effort to improve on this situation was made in 1856, when a man was towed in a sort of kite which was hauled by a vehicle moving on the ground. In February of the present year, Lieut. John Rodgers, U.S.N., was lifted 400 feet from the deck of the cruiser *Pennsylvania* by a train of eleven large kites, the vessel steaming at twelve knots against an eight-knot breeze. The aviator made observations and took photographs for about fifteen minutes, while suspended from a tail cable about 100 feet astern. In the absence of a sufficient natural breeze, an artificial wind was thus produced by the motion imparted to the kite; and the device permitted of reaching some destination. The next step was obviously to get rid of the tractive vehicle and tow rope by carrying propelling machinery on the kite. This had been accomplished by Langley in 1896, who flew a thirty-pound model nearly a mile, using a steam engine for power. The gasoline engine, first employed by Santos-Dumont (in a dirigible balloon) in 1901, has made possible the present day *aeroplane*.

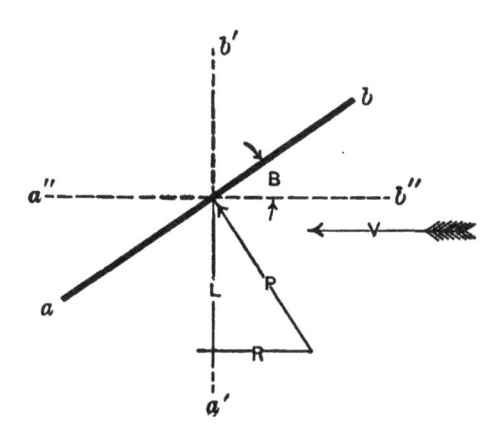

What "keeps it up," in the case of this device, is likewise its velocity. Looking from the side, *ab* is the sail of the aeroplane, which is moving toward the right at such speed as to produce the equivalent of an air velocity V to the left. This velocity causes the direct pressure P, equivalent to a lifting force L and a retarding force R. The latter is the force which must be overcome by the motor: the former must suffice to overcome the whole weight of the apparatus. Travel in an aeroplane is like skating rapidly over very thin ice: the air literally "doesn't have time to get away from underneath."

If we designate the angle made by the wings (ab) with the horizontal (V) as B, then P increases as B increases,

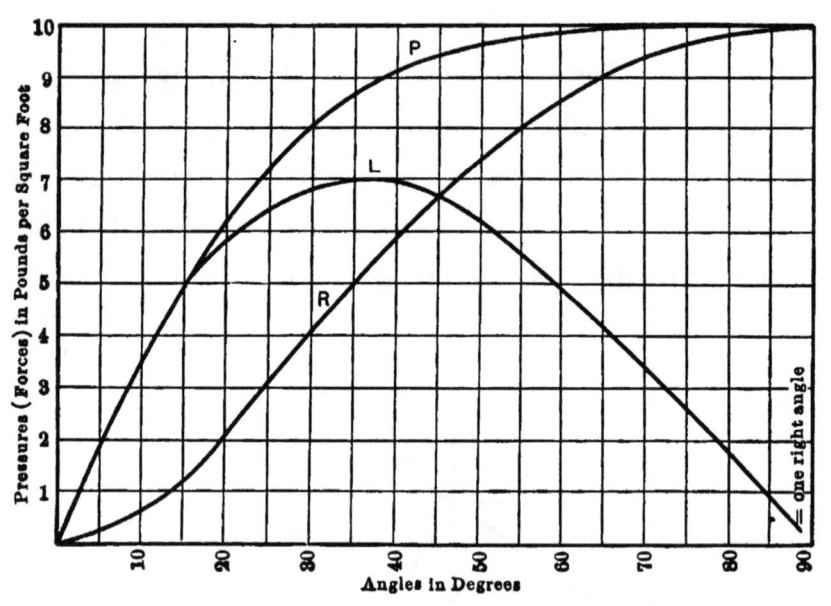

DIRECT, LIFTING, AND RESISTING FORCES

If the pressure is 10 lbs. when the wind blows directly toward the surface (at an angle of 90 degrees), then the forces for other angles of direction are as shown on the diagram. The *amounts* of all forces depend upon the wind velocity: that assumed in drawing the diagram was about 55 miles per hour. But the *relations* of the forces are the same for the various angles, no matter what the velocity.

while (as has been stated) the ratio of L to R decreases. When the angle B is a right angle, the wings being in the position a'b', P has its maximum value for direct wind — $\frac{1}{300}$ of the square of the velocity, in pounds per square foot; but L is zero and R is equal to P. The plane would have no lifting power. When the angle B becomes zero, position $a''b''$, wings being horizontal, P becomes zero and (so far as we can now judge) the plane has neither lifting power nor retarding force. At some intermediate position, like ab, there will be appreciable lifting and retarding forces. The chart shows the approximate lifting force, in pounds per square foot, for various angles. This force becomes a maximum at an angle of 45° (half a right angle). We are not yet prepared to consider why in all actual aeroplanes the angle of inclination is much less than this. The reason will be shown presently. At this stage of the discussion we may note that the lifting power per square foot of sail area varies with

    the square of the velocity, *and*
    the angle of inclination.

The total lifting power of the whole plane will also vary with its area. As we do not wish this whole lifting power to be consumed in overcoming the dead weight of the machine itself, we must keep the parts light, and in particular must use for the wings a fabric of light weight per unit of surface. These fabrics are frequently the same as those used for the envelopes of balloons.

Since the total supporting power varies both with the sail area and with the velocity, we may attain a given capacity either by employing large sails or by using high speed. The size of sails for a given machine varies inversely as the square of the speed. The original Wright machine had 500 square feet of wings and a speed of forty miles per hour. At eighty miles per hour the necessary sail area for this machine would be only 125 square feet; and at 160 miles per hour it would be only $31\frac{1}{4}$ square feet: while if we attempted to run the machine at ten miles per hour we should need a sail area of 8000 square feet. This explains why the aeroplane cannot go slowly.

It would seem as if when two or more superposed sails were used, as in biplanes, the full effect of the air would not be realized, one sail becalming the other. Experiments have shown this to be the case; but there is no great reduction in lifting power unless the distance apart is considerably less than the width of the planes.

In all present aeroplanes the sails are concaved on the under side. This serves to keep the air from escaping from underneath as rapidly as it otherwise would, and increases the lifting power from one-fourth to one-half over that given by our $\frac{1}{300}$ rule: the divisor becoming roughly about 230 instead of 300.

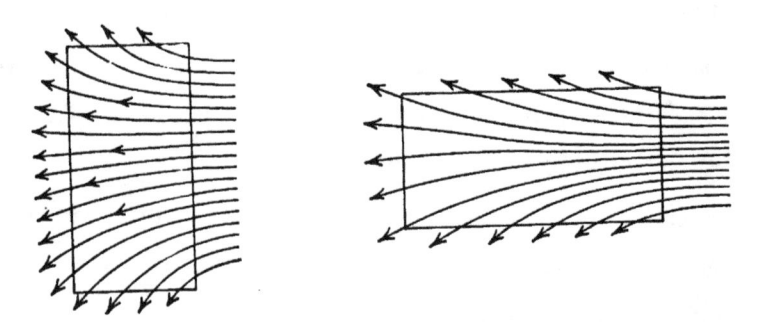

Why are the wings placed crosswise of the machine, when the other arrangement — the greatest dimension in the line of flight — would seem to be stronger? This is also done in order to "keep the air from escaping from underneath." The sketch shows how much less easily the air will get away from below a wing of the bird-like spread-out form than from one relatively long and narrow but of the same area.

A sustaining force of two pounds per square foot of area has been common in ordinary aeroplanes and is perhaps comparable with the results of bird studies: but this figure is steadily increasing as velocities increase.

WHY SO MANY SAILS?

Thus far a single wing or pair of wings would seem to fully answer for practicable flight: yet every actual aeroplane has several small wings at various points. The necessity for one of these had already been discovered in the kite, which is built with a balancing tail. In the sketch on page 18 it appears that the particles of air which are near the upper edge of the surface are more obstructed in their effort to get around and past than those near the lower edge. They have to turn almost completely about, while the others are merely deflected. This means that on the whole the upper air particles will exert more pressure than the lower particles and that the "center of pressure" (the point where the entire force of the wind may be assumed to act) will be, not at the center of the surface, but at a point some distance *above* this center. This action is described as the "displacement of the center of pressure." It is known that the displacement is greatest for least inclinations of surface (as might be surmised from the sketch already referred to), and that it is always proportional to the dimension of the surface in the direction of movement; *i.e.*, to the length of the line *ab*.

If the weight W of the aeroplane acts downward at the center of the wing (at *o* in the accompanying sketch), while the direct pressure P acts at some point *c* farther along toward the upper edge of the wing, the two forces W and P tend to revolve the whole wing in the direction indicated by the curved arrow. This rotation, in an aero-

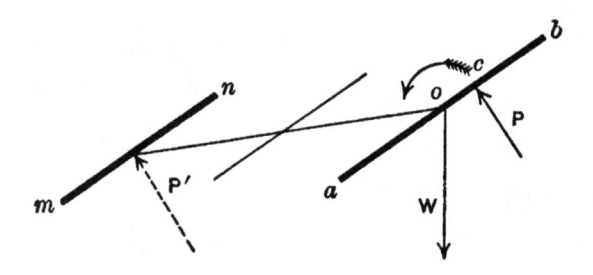

plane, is resisted by the use of a tail plane or planes, such as *mn*. The velocity produces a direct pressure P' on the tail plane, which opposes, like a lever, any rotation due to the action of P. It may be considered a matter of rather nice calculation to get the area and location of the tail plane just right: but we must remember that the amount of pressure P' can be greatly varied by changing the inclination of the surface *mn*. This change of inclination is effected by the operator, who has access to wires which are attached to the pivoted tail plane. It is of course permissible to place the tail plane *in front* of the main planes —

as in the original Wright machine illustrated: but in this case, with the relative positions of W and P already shown, the forward edge of the tail plane would have to be depressed instead of elevated. The illustration shows the tail built as a biplane, just as are the principal wings (page 141).

Suppose the machine to be started with the tail plane in a horizontal position. As its speed increases, it rises and at the same time (if the weight is suspended from the center of the main planes) tilts backward. The tilting can be stopped by swinging the tail plane on its pivot so as to oppose the rotative tendency. If this control is not carried too far, the main planes will be allowed to maintain some of their excessive inclination and ascent will continue. When the desired altitude has been attained, the inclination of the main planes will, by further swinging of the tail plane, be reduced to the normal amount, at which the supporting power is precisely equal to the load; and the machine will be in vertical equilibrium: an equilibrium which demands at every moment, however, the attention of the operator.

In many machines, ascent and tilting are separately controlled by using two sets of transverse planes, one set placed forward, and the other set aft, of the main planes. In any case, quick ascent can be produced only by an increase in the lifting force L (see sketch, page 24) of the main planes: and this force is increased by enlarging the angle of inclination of the main planes, that is, by a controlled and partial tilting. The forward transverse wing which produces this tilting is therefore called the *elevating rudder* or elevating plane. The rear transverse plane which checks the tilting and steadies the machine is often

ROE'S TRIPLANE AT WEMBLEY
(From Brewer's *Art of Aviation*)

described as the *stabilizing plane*. *Descent* is of course produced by *decreasing* the angle of inclination of the main planes.

STEERING

If we need extra sails for stability and ascent or descent, we need them also for changes of horizontal direction. Let ab be the top view of the main plane of a machine, following the course xy. At rs is a vertical plane called the *steering rudder*. This is pivoted, and controlled by the operator by means of the wires t, u. Let the rudder be suddenly shifted to the position $r's'$. It will then be subjected to a pressure P' which will swing the whole machine into the new position shown by the dotted lines, its course becoming $x'y'$. The steering rudder may of course be double, forming a vertical biplane, as in the Wright machine shown below.

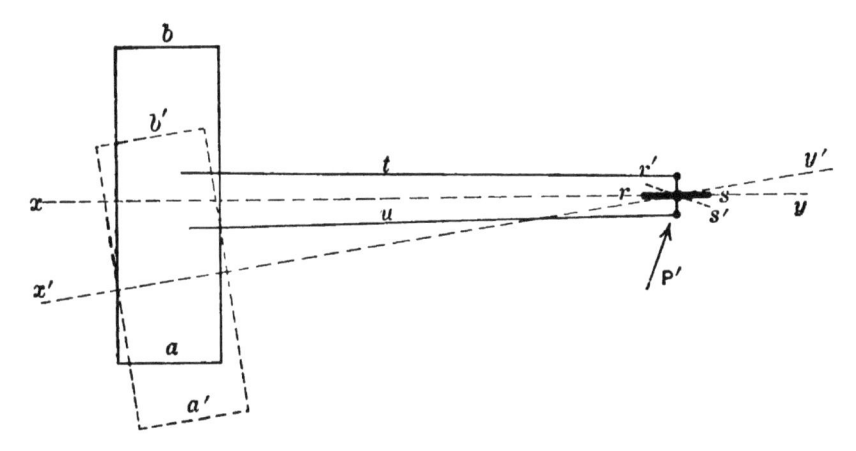

Successful steering necessitates lateral resistance to drift, *i.e.*, a fulcrum. This is provided, to some extent, by

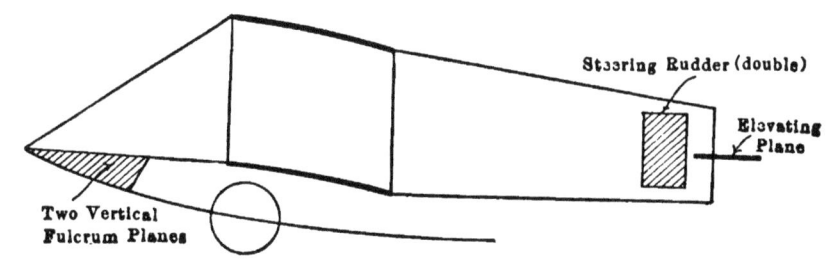

RECENT TYPE OF WRIGHT BIPLANE

the stays and frame of the machine; and in a much more ample way by the vertical planes of the original Voisin cellular biplane. A recent Wright machine had vertical planes forward probably intended for this purpose.

It now begins to appear that the aviator has a great many things to look after. There are many more things requiring his attention than have yet been suggested. No one has any business to attempt flying unless he is superlatively cool-headed and has the happy faculty of instinctively doing the right thing in an emergency. Give a chauffeur a high power automobile running at maximum speed on a rough and unfamiliar road, and you have some conception of the position of the operator of an aeroplane. It is perhaps not too much to say that to make the two positions fairly comparable we should *blindfold* the chauffeur.

Broadly speaking, designers may be classed in one of two groups — those who, like the Wrights, believe in training the aviator so as to qualify him to properly handle his complicated machine; and those who aim to simplify the whole question of control so that to acquire the necessary ability will not be impossible for the average man. If aviation is to become a popular sport, the latter ideal must prevail. The machines must be more automatic and the aviator must have time to enjoy the scenery. In France, where amateur aviation is of some importance, progress has already been made in this direction. The universal steering head, for example, which not only revolves like that of an automobile, but is hinged to permit of additional movements, provides for simultaneous control of the steering rudder and the main plane warping, while scarcely demanding the conscious thought of the operator.

TURNING CORNERS

A YEAR elapsed after the first successful flight at Kitty Hawk before the aviator became able to describe a circle in the air. A later date, 1907, is recorded for the first European half-circular flight: and the first complete circuit, on the other side of the water, was made a year after that; by both biplane and monoplane. It was in the same year that Louis Blériot made the pioneer cross-country trip of twenty-one miles, stopping at will *en route* and returning to his starting point.

WHAT HAPPENS WHEN MAKING A TURN

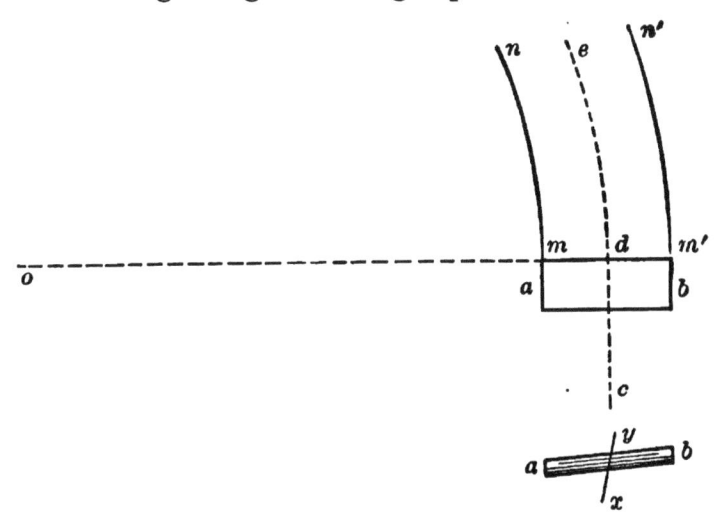

We are looking downward on an aeroplane *ab* which has been moving along the straight path *cd*. At *d* it begins to describe the circle *de*, the radius of which is *od*, around the center *o*. The outer portion of the plane, at the edge *b*, must then move faster than the inner edge *a*. We have seen that the direct air pressure on the plane is proportional to the square of the velocity. The direct pressure *P* (see sketch on page 22) will then be greater at the outer than at the inner limb; the lifting force *L* will also be greater and the outer limb will tend to rise, so that the plane (viewed from the rear) will take the inclined position shown in the lower view: and this inclination will increase as long as the outer limb travels faster than the inner limb; that is, as long as the orbit continues to be curved. Very soon, then, the plane will be completely tipped over.

Necessarily, the two velocities have the ratio $om:om'$; the respective lifting forces must then be proportional to the squares of these distances. The difference of lifting forces, and the tendency to overturn, will be more important as the distances most greatly differ: which is the case when the distance *om* is small as compared with *mm'*. The shorter the radius of curvature, the more dangerous, for a given machine, is a circling flight: and in rounding a curve of given radius the most danger is attached to the machine of greatest spread of wing.

Lateral Stability

This particular difficulty has considerably delayed the development of the aeroplane. It may, however, be overcome by very simple methods — simple, at least as far as their mechanical features are concerned. If the outer limb of the plane is tilted upward, it is because the wind pressure is greater there. The wind pressure is greater because the velocity is greater. We have only to increase the wind pressure at the inner limb, in order to restore equilibrium. This cannot be done by adjusting the velocity, because the velocity is fixed by the curvature of path required: but the total wind pressure depends upon the *sail area* as well as the velocity; so that by increasing the surface at the inner limb we may equalize the value of L, the lifting force, at the two ends of the plane. This increase of surface must be a temporary affair, to be discontinued when moving along a straight course.

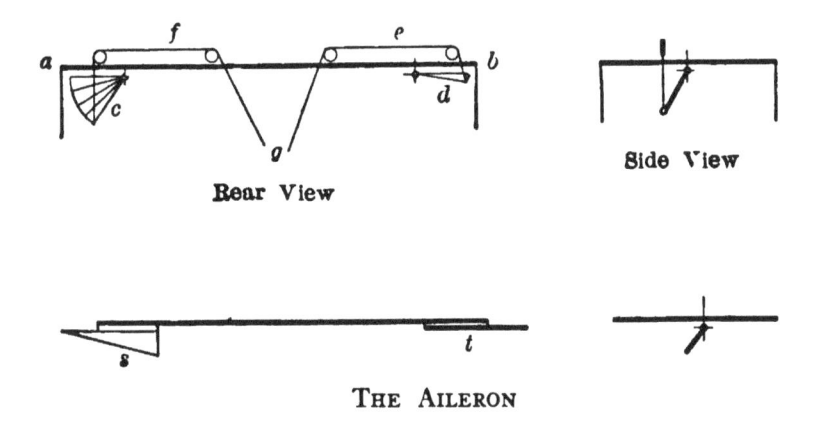

THE AILERON

Let us stand in the rear of an aeroplane, the main wing of which is represented by ab. Let the small fan-shaped wings c and d be attached near the ends, and let the control wires, e, f, passing to the operator at g, be employed to close and unclasp the fans. If these fans are given a forward inclination at the top, as indicated in the end view, they will when spread out exert an extra lifting force. A fan will be placed at each end. They will be ordinarily folded up: but when rounding a curve the aviator will open the fan on the inner or more slowly moving limb of the main plane. This represents one of the first forms of the *aileron* or wing-tip for lateral control.

The more common present form of aileron is that shown in the lower sketch, at *s* and *t*. The method of control is the same.

The cellular Voisin biplanes illustrate an attempt at self-sufficing control, without the interposition of the aviator. Between the upper and lower sails of the machine there were fore and aft vertical partitions. The idea was that when the machine started to revolve, the velocity of rotation would produce a pressure against these partitions

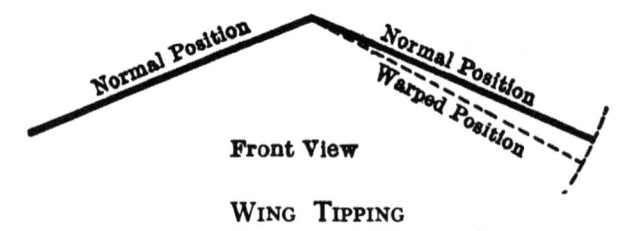

Front View

WING TIPPING

which would obstruct the tipping. But rotation may take place slowly, so as to produce an insufficient pressure for control, and yet be amply sufficient to wreck the apparatus. The use of extra vertical rudder planes, hinged on a horizontal longitudinal axis, is open to the same objection.

WING WARPING

In some monoplanes with the inverted *V* wing arrangement, a dipping of one wing answers, so to speak, to increase its concavity and thus to augment the lifting force on that side. The sketch shows the normal and distorted arrangement of wings: the inner limb being the one bent down in rounding a curve. An equivalent plan was to change the angle of inclination of one-half the sail by swinging it about a horizontal pivot at the center or at the rear edge: some machines have been built with sails divided in the center. The obvious objection to both of these plans is that too much mechanism is necessary in order to distort what amounts to nearly half the whole machine. They remind one of Charles Lamb's story of the discovery of roast pig.

The distinctive feature of the Wright machines lies in

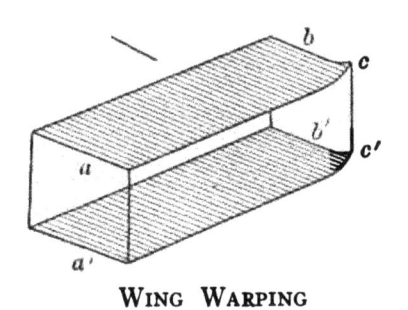

WING WARPING

the warping or distorting of the *ends only* of the main planes. This is made possible, not by hinging the wings in halves, but by the flexibility of the framework, which is sufficiently pliable to permit of a considerable bending without danger. The operator, by pulling on a stout wire linkage, may tip up (or down) the corners cc' of the sails at one limb, thus decreasing or increasing the effective surface acted on by the wind, as the case may require.

The only objection is that the scheme provides one more thing for the aviator to think about and manipulate.

Automatic Control

Let us consider again the condition of things when rounding a curve, as in the sketch on page 32. As long as the machine is moving forward in a straight line, the operator sits upright. When it begins to tip, he will unconsciously tip himself the other way, as represented by the line xy in the rear view. Any bicyclist will recognize this as plausible. Why not take advantage of this involuntary movement to provide a stabilizing force? If operating wires are attached to the aviator's belt and from thence connected with ailerons or wing-warping devices, then by a proper proportioning of levers and surfaces to the probable swaying of the man, the control may become automatic. The idea is not new; it has even been made the subject of a patent.

The Gyroscope

This device for automatic control is being steadily developed and may ultimately supersede all others. It uses the inertia of a fast-moving fly wheel for control, in a manner not unlike that contemplated in proposed methods of automatic balancing by the action of a suspended pendulum. Every one has seen the toy gyroscope and perhaps has wondered at its mysterious ways. The mathematical analysis of its action fills volumes: but some idea of what it does, and why, may perhaps be gathered at the expense of a very small amount of careful attention. The wheel *acbd*, a thin disc, is spinning rapidly about the axle *o*. In the side view, *ab* shows the edge of the wheel, and *oo'* the

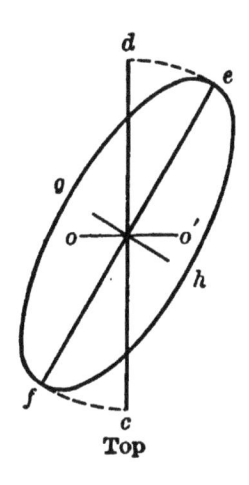

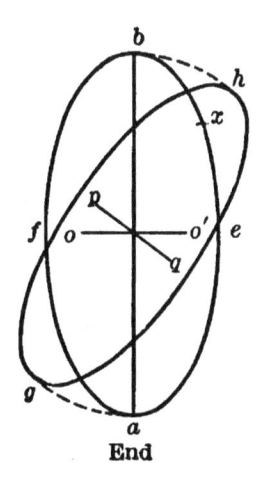

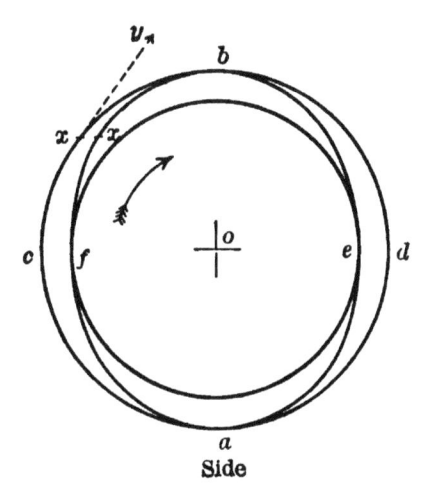

THE GYROSCOPE

axle. This axle is not fixed, but may be conceived as held in some one's fingers. Now suppose the right-hand end of the axle (*o'*) to be suddenly moved toward us (away from the paper) and the left-hand (*o*) to be moved away.

The wheel will now appear in both views as an ellipse, and it has been so represented, as *afbe*. Now, any particle, like x, on the rim of the wheel, will have been regularly moving in the circular orbit *cb*. The tendency of any body in motion is to move indefinitely in a straight line. The cohesion of the metal of the disc prevents the particle x from flying off at a straight line tangent, *xy*, and it is constrained, therefore, to move in a circular orbit. Unless some additional constraint is imposed, it will at least remain in this orbit and will try to remain in its plane of rotation. When the disc is tipped, the plane of rotation is changed, and the particle is required, instead of (so to speak) remaining in the plane of the paper — in the side view — to approach and pass through that plane at *b* and afterward to continue receding from us. Under ordinary circumstances, this is just what it would do: but if, as in the gyroscope, the axle oo' is perfectly free to move in any direction, the particle x will refuse to change its direction of rotation. Its position has been shifted: it no longer lies in the plane of the paper: but it will at least persist in rotating in a parallel plane: and this persistence forces the revolving disc to swing into the new position indicated by the curve *hg*, the axis being tipped into the position *pq*. The whole effect of all particles like x in the entire wheel will be found to produce precisely this condition of things: if we undertake to change the plane of rotation by shifting the axle in a horizontal plane, the device itself will (if not prevented) make a further change in the plane of rotation by shifting the axle in a vertical plane.

A revolving disc mounted on the gyroscopic framework therefore resists influences tending to change its plane of rotation. If the device is placed on a steamship, so that when the vessel rolls a change of rotative plane is produced, the action of the gyroscope will resist the rolling tendency of the vessel. All that is necessary is to have the wheel revolving in a fore and aft plane on the center line of the vessel, the axle being transverse and firmly attached to the vessel itself. A small amount of power (consumed in revolving the wheel) gives a marked steadying effect. The same location and arrangement on an aeroplane will suffice to overcome tendencies to transverse rotation when rounding curves. The device itself is automatic, and requires no attention, but it does unfortunately require power to drive it and it adds some weight.

The gyroscope is being tested at the present time on some of the aeroplanes at the temporary army camps near San Antonio, Texas.

Wind Gusts

This feature of aeronautics is particularly important, because any device which will give automatic stability when turning corners will go far toward making aviation a safe amusement. Inequalities of velocity exist not only on curves, but also when the wind is blowing at anything but uniform velocity across the whole front of the machine.

The slightest "flaw" in the wind means an at least temporary variation in lifting force of the two arms. Here is a pregnant source of danger, and one which cannot be left for the aviator to meet by conscious thought and action. It is this, then, that blindfolds him: he cannot see the wind conditions in advance. The conditions are upon him, and may have done their destructive work, before he can prepare to control them. We must now study what these conditions are and what their influence may be on various forms of aerial navigation: after which, a return to our present subject will be possible.

AIR AND THE WIND

THE air that surrounds us weighs about one-thirteenth of a pound per cubic foot and exerts a pressure, at sea level, of nearly fifteen pounds per square inch. Its temperature varies from 30° below to 100° above the Fahrenheit zero. The pressure of the air decreases about one-half pound for each thousand feet of altitude; at the top of Mt. Blanc it would be, therefore, only about six pounds per square inch. The temperature also decreases with the altitude. The weight of a cubic foot, or *density*, which, as has been stated, is one-thirteenth of a pound ordinarily, varies with the pressure and with the temperature. The variation with pressure may be described by saying that the *quotient* of the pressure by the density is constant: one varies in the same ratio as the other. Thus, at the top of Mt. Blanc (if the temperature were the same as at sea level), the density of air would be about $\frac{6}{15} \times \frac{1}{13} = \frac{2}{65}$: less than half what it is at sea level. As to temperature, if we call our Fahrenheit zero 460°, and correspondingly describe other temperatures — for instance, say that water boils at 672° — then (pressure being unchanged) the *product* of the density and the temperature is constant. If the density at sea level and zero temperature is one-thirteenth pound, then that at sea level and 460° Fahrenheit would be

$$\frac{0+460}{460+460} \times \tfrac{1}{13} = \tfrac{1}{26}.$$

These relations are particularly important in the design of all balloons, and in computations relating to aeroplane flight at high altitudes. We shall be prepared to appreciate some of their applications presently.

Generally speaking, the atmosphere is always in motion, and moving air is called wind. Our meteorologists first studied winds near the surface of the ground: it is only of late years that high altitude measurements have been considered practically desirable. Now, records are obtained by the aid of kites up to a height of nearly four miles: estimates of cloud movements have given data on wind velocities at heights above six miles: and much greater heights have been obtained by free balloons equipped with instruments for recording temperatures, pressures, altitude, time, and other data.

When the Eiffel Tower was completed, it was found that the average wind velocity at its summit was about four times that at the base. Since that time, much attention has been given to the contrasting conditions of surface and upper breezes as to direction and velocity.

Air is easily impeded in its movement, and the well-known uncertainties of the weather are closely related to local variations in atmospheric pressure and temperature. When near the surface of the ground, impingement against irregularities therein — hills, cliffs, and buildings — makes the atmospheric currents turbulent and irregular. Where

Air and the Wind

there are no surface irregularities, as on a smooth plain or over water, the friction of the air particles passing over the surface still results in a stratification of velocities. Even on a mountain top, the direction and speed of the wind are less steady than in the open where measured by a captive balloon. The stronger the wind, the greater, relatively, is the irregularity produced by surface conditions. Further, the earth's surface and its features form a

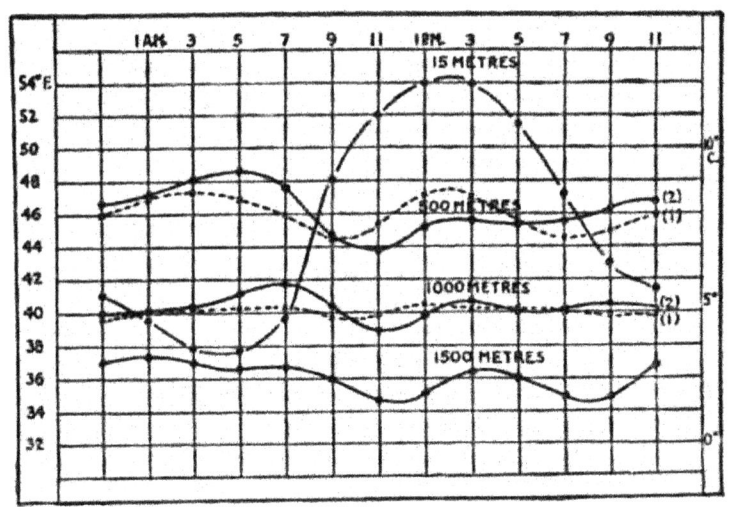

DIURNAL TEMPERATURES AT DIFFERENT HEIGHTS
(From Rotch's *The Conquest of the Air*)

vast sponge for sun heat, which they transfer in turn to the air in an irregular way, producing those convectional currents peculiar to low altitudes, the upper limit of which is marked by the elevation of the cumulus clouds. Near the surface, therefore, wind velocities are lowest in the early morning, rising to a maximum in the afternoon.

Every locality has its so-called "prevailing winds." Considering the compass as having eight points, one of those points may describe as many as 40% of all the winds at a given place. The direction of prevalence varies with the season. The range of wind velocities is also a matter of local peculiarity. In Paris, the wind speed exceeds thirty-four miles per hour on only sixty-eight days in the average year, and exceeds fifty-four miles on only fifteen days. Observations at Boston show that the velocity of the wind exceeds twenty miles per hour on half the days in winter and on only one-sixth the days in summer. Our largest present dirigible balloons have independent speeds of about thirty-four miles per hour and are therefore available (at some degree of effectiveness) for nearly ten months of the year, in the vicinity of Paris. In a region of low wind velocities — like western Washington, in this country — they would be available a much greater proportion of the time. To make the dirigible able to at least move nearly every day in the average year — in Paris — it must be given a speed of about fifty-five miles per hour.

Figures as to wind velocity mean little to one unaccustomed to using them. A five-mile breeze is just "pleasant." Twelve miles means a brisk gale. Thirty miles is a high wind: fifty miles a serious storm (these are the winds the aviator constantly meets): one hundred miles is perhaps about the maximum hurricane velocity.

As we ascend from the surface of the earth, the wind velocity steadily increases; and the excess velocity of winter winds over summer winds is as steadily augmented. Thus, Professor Rotch found the following variations:

Altitude in Feet	Annual Average Wind Velocity, Feet per Second
656	23.15
1,800	32.10
3,280	35.
8,190	41.
11,440	50.8
17,680	81.7
20,970	89.
31,100	117.5

Altitude in Feet	Average Wind Velocities, Feet per Second	
	Summer	Winter
656 to 3,280	24.55	28.80
3,280 to 9,810	26.85	48.17
9,810 to 16,400	34.65	71.00
16,400 to 22,950	62.60	161.5
22,950 to 29,500	77.00	177.0

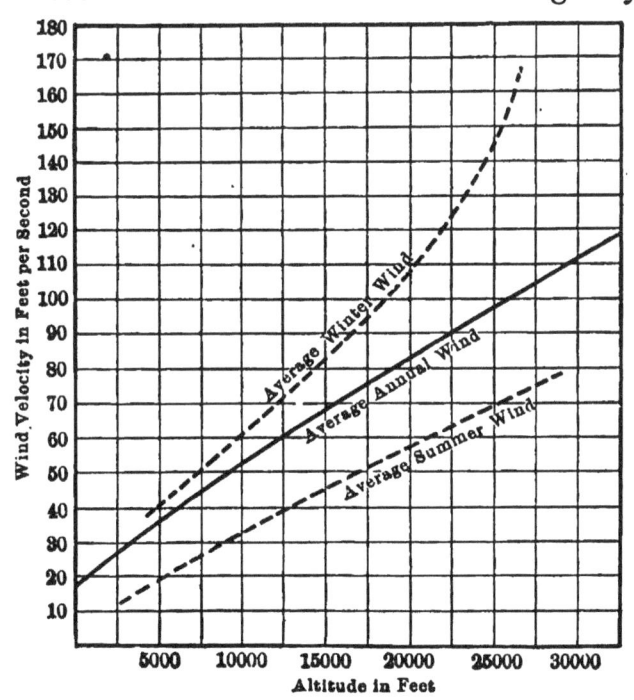

These results are shown in a more striking way by the chart. At a five or six mile height, double-barreled hurricanes at speeds exceeding 200 miles per hour are not merely possible; they are part of the regular order of things, during the winter months.

The winds of the upper air, though vastly more powerful, are far less irregular than those near the surface: and the directions of prevailing winds are changed. If 50% of the winds, at a given location on the surface, are from the southwest, then at as moderate an elevation as even 1000 feet, the prevailing direction will cease to be from southwest; it may become from west-southwest; and the proportion of total winds coming from this direction will not be 50%. These factors are represented in meteorological papers by what is known as the *wind rose*. From the samples shown, we may note that 40% of the surface winds at Mount Weather are from the northwest; while at some elevation not stated the most prevalent of the winds (22% of the total) are westerly. The direction of prevalence has changed through one-eighth of the possible circle, and in a *counter-clockwise* direction. This is contrary to the usual variation described by the so-called Broun's Law, which asserts that as we ascend the direction of prevalence rotates around the circle like the hands of a watch; being, say, from northwest at the surface, from north at some elevation, from northeast at a still higher elevation, and so on. At a great height, the change in direction may become total: that is, the high altitude winds blow in the exactly opposite direction to that of the surface winds. In the temperate regions, most of the high altitude winds are from the west: in the tropics,

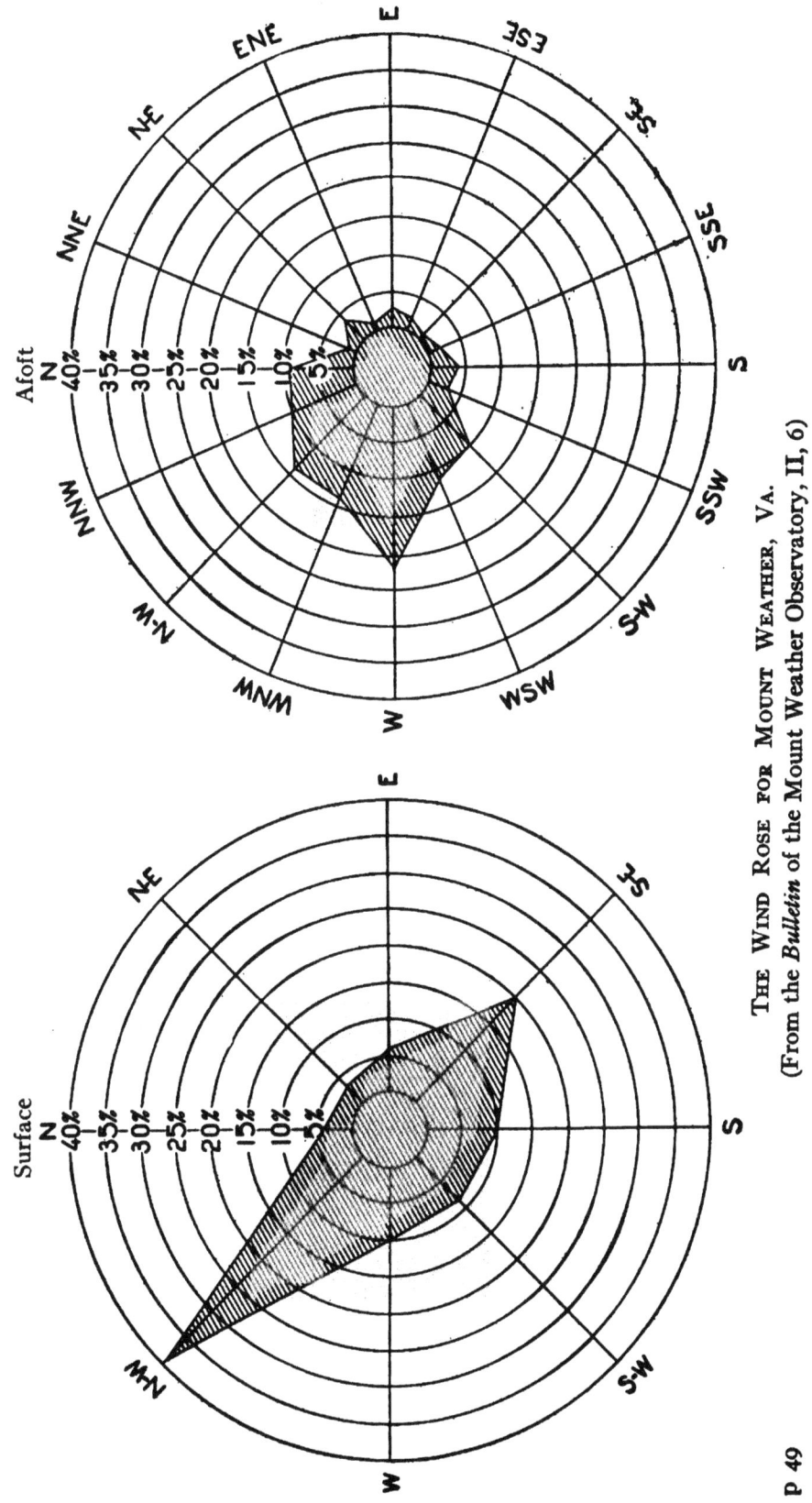

THE WIND ROSE FOR MOUNT WEATHER, VA.
(From the *Bulletin* of the Mount Weather Observatory, II, 6)

the surface winds blow *toward* the west and toward the equator; being northeasterly in the northern hemisphere and southeasterly in the southern: and there are undoubtedly equally prevalent high-altitude counter-trades.

The best flying height for an aeroplane over a flat field out in the country is perhaps quite low — 200 or 300 feet: but for cross-country trips, where hills, rivers, and buildings disturb the air currents, a much higher elevation is necessary; perhaps 2000 or 3000 feet, but in no case more than a mile. The same altitude is suitable for dirigible balloons. At these elevations we have the conditions of reasonable warmth, dryness, and moderate wind velocities.

Sailing Balloons

In classifying air craft, the sailing balloon was mentioned as a type intermediate between the drifting balloon and the dirigible. No such type has before been recognized: but it may prove to have its field, just as the sailing vessel on the sea has bridged the gap between the raft and the steamship. It is true that tacking is impossible, so that our sailing balloons must always run before the wind: but they possess this great advantage over marine sailing craft, that by varying their altitude they may always be able to find a favorable wind. This implies adequate altitude control, which is one of the problems not yet solved for lighter-than-air flying machines: but when it has been solved we shall go far toward attaining a dirigible balloon without motor or propeller; a true sailing craft.

Air and the Wind

This means more study and careful utilization of stratified atmospheric currents. Professor Rotch suggests the utilization of the upper westerly wind drift across the American continent and the Atlantic Ocean, which would carry a balloon from San Francisco to southern Europe at

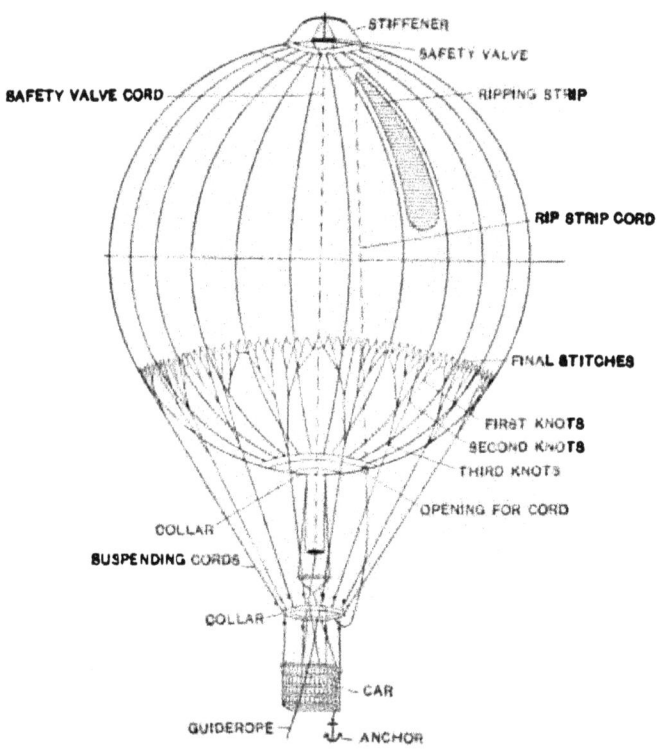

DIAGRAM OF PARTS OF A DRIFTING BALLOON

a speed of about fifty feet per second — thirty-four miles per hour. Then by transporting the balloon to northern Africa, the northeast surface trade wind would drive it back to the West Indies at twenty-five miles per hour. This without any motive power: and since present day dirigibles are all short of motive power for complete dirigibility, we must either make them much more powerful or else adopt the sailing principle, which will permit of

GLIDDEN AND STEVENS GETTING AWAY IN THE "BOSTON"
(Leo Stevens, N.Y.)

actually decreasing present sizes of motors, or even possibly of omitting them altogether. Our next study is, then, logically, one of altitude control in balloons.

Air and the Wind

Field and Speed

An *aerostat* (non-dirigible balloon), unless anchored, drifts at the speed of the wind. To the occupants, it seems to stand still, while the surface of the earth below appears to move in a direction opposite to that of the wind. In the sketch, if the independent velocity of a *dirigible* balloon

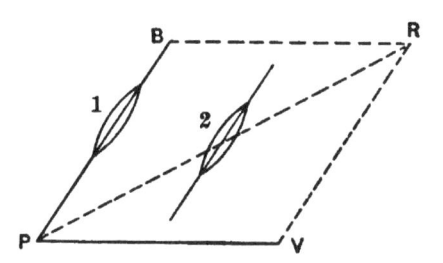

be PB, the wind velocity PV, then the actual course pursued is PR, although the balloon always points in the direction PB, as shown at 1 and 2. If the speed of the wind exceed that of the balloon, there will be some directions in which the latter cannot progress. Thus, let PV be the

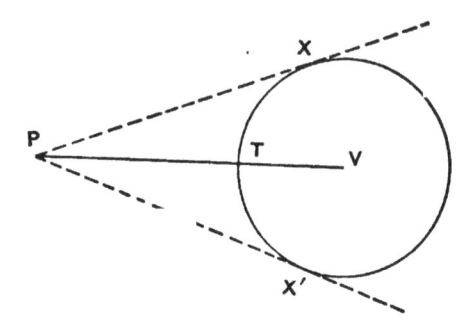

wind velocity and TV the independent speed of the balloon. The tangents PX, PX', include the whole "field of action" possible. The wind direction may change during flight, so that the initial objective point may become unattainable, or an initially unattainable point may be brought within the field. The present need is to increase independent speeds from thirty or forty to fifty or sixty miles per hour, so that the balloon will be truly dirigible (even if at low effectiveness) during practically the whole year.

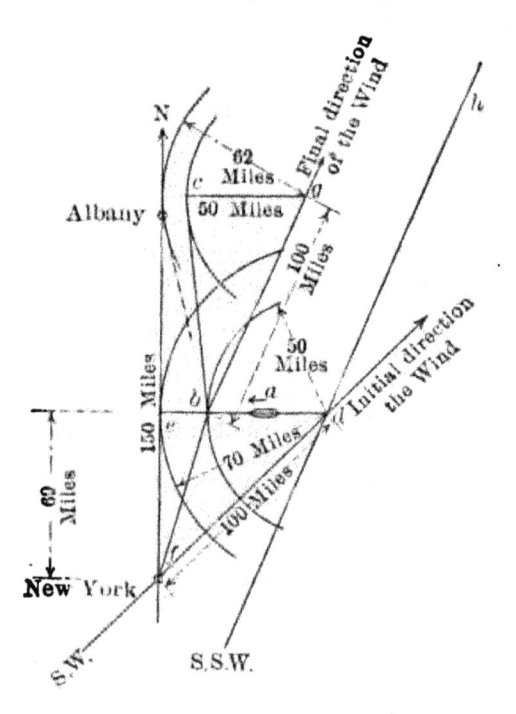

Suppose a dirigible to start on a trip from New York to Albany, 150 miles away. Let the wind be a twenty-five mile breeze from the southwest. The wind alone tends to carry the balloon from New York to the point *d* in four hours. If the balloon meanwhile be headed due west, it would need an independent velocity of its own having the same ratio to that of the wind as that of *de* to *fd*, or about seventeen and one-half miles per hour. Suppose its independent speed to be only twelve and one-half miles; then after four hours it will be at the position b, assuming it to have been continually headed due west, as indicated

COUNT ZEPPELIN

at a. It will have traveled northward the distance fe, apparently about sixty-nine miles.

After this four hours of flight, the wind suddenly changes to south-southwest. It now tends to carry the balloon to *g* in the next four hours. Meanwhile the balloon, heading west, overcomes the easterly drift, and the balloon actually lands at *c*. Unless there is some further favorable shift of the wind it cannot reach Albany. If, during the second four hours, its independent speed could have been increased to about fifteen and a half miles it would have just made it. The actual course has been *fbc:* a drifting balloon would have followed the course *fdh, dh* being a course parallel to *bg*.

GAS AND BALLAST

A CUBICAL block of wood measuring twelve inches on a side floats on water because it is lighter than water; it weighs, if yellow pine, thirty-eight pounds, whereas the same volume of water weighs about sixty-two pounds. Any substance weighing more than sixty-two pounds to the cubic foot would sink in water.

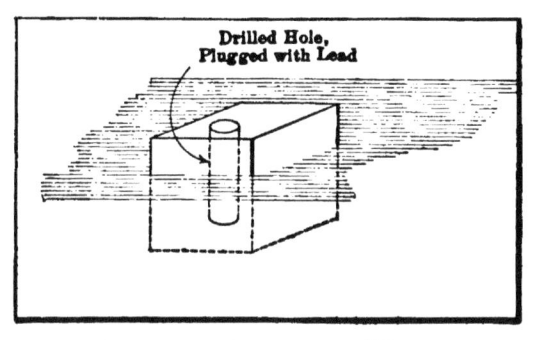

BUOYANT POWER OF WOOD

If our block of wood be drilled, and *lead* poured in the hole, the total size of wood-and-lead block being kept constantly at one cubic foot, the block will sink as soon as its whole weight exceeds sixty-two pounds. Ignoring the wood removed by boring (as, compared with the lead which replaces it, an insignificant amount), the weight of lead plugged in may reach twenty-four pounds before the block will sink.

This figure, twenty-four pounds, the difference between

sixty-two and thirty-eight pounds, then represents the maximum buoyant power of a cubic foot of wood in water. It is the difference between the weight of the wood block and the weight of the water it displaces. If any weight

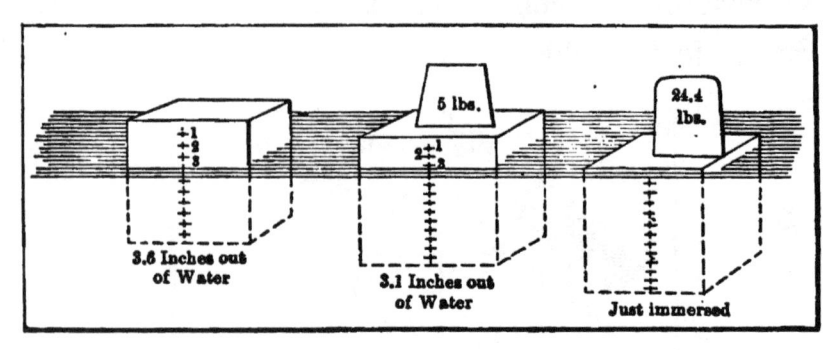

ONE CUBIC FOOT OF WOOD LOADED IN WATER

less than this is added to that of the wood, the block will float, projecting above the water's surface more or less, according to the amount of weight buoyed up. It will not rise entirely from the water, because to do this it would need to be lighter, not only than water, but than air.

BUOYANCY IN AIR

There are *gases*, if not woods, lighter than air: among them, coal gas and hydrogen. A "bubble" of any of these gases, if isolated from the surrounding atmosphere, cannot sink but must rise. At the same pressure and temperature, hydrogen weighs about one-fifteenth as much as air; coal gas, about one-third as much. If a bubble of either of these gases be isolated in the atmosphere, it must continually rise, just as wood immersed in water will rise when liberated. But the wood will stop when it reaches the surface of the water, while there is no reason to suppose that the hydrogen or coal gas bubbles will ever stop. The hydrogen bubble can be made to remain stationary if it is weighted down with something of about fourteen times its own weight (thirteen and one-half times, accurately). Perhaps it would be better to say that it would still continue to rise

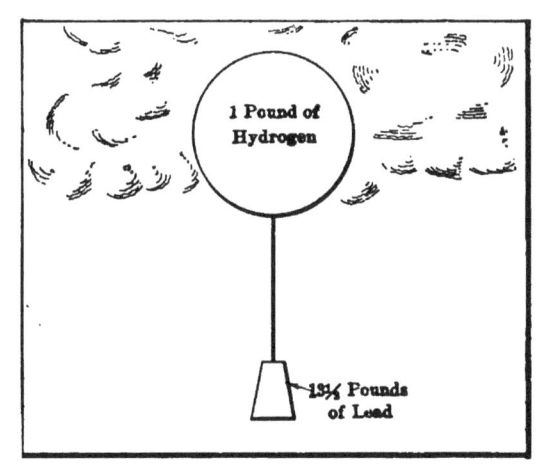

BUOYANT POWER OF HYDROGEN

slowly because that additional something would itself displace some additional air; but if the added weight is a solid body, its own buoyancy in air is negligible.

Our first principle is, then, that at the same pressure and temperature, any gas lighter than air, if properly confined, will exert a net lifting power of $(n-1)$ times its own weight, where n is the ratio of weights of air and gas per cubic foot.

If the pressures and temperatures are different, this principle is modified. In a balloon, the gas is under a

LEBAUDY'S "JAUNE"

pressure slightly in excess of that of the external atmosphere: this decreases its lifting power, because the weight of a given volume of gas is greater as the pressure to which it is subjected is increased. The weight of a given volume we have called the *density:* and, as has been stated, if the temperature be unchanged, the density varies directly as the pressure.

The pressure in a balloon is only about 1% greater than that of the atmosphere at sea level, so that this factor has only a slight influence on the lifting power. That it leads to certain difficulties in economy of gas will, however, soon be seen.

The temperature of the gas in a balloon, one might think, would naturally be the same as that of the air outside: but the surface of the balloon envelope has an absorbing capacity for heat, and on a bright sunny day the gas may be considerably warmed thereby. This action increases the lifting power, since increase of temperature (the pressure remaining fixed) decreases the density of a gas. To avoid this possibly objectionable increase in lifting power, balloons are sometimes painted with a nonabsorbent color. One of the first Lebaudy balloons received a popular nickname in Paris on account of the yellow hue of its envelope.

Suppose we wish a balloon to carry a total weight, including that of the envelope itself, of a ton. If of hydrogen, it will have to contain one fifteenth of this weight or about 133 pounds of that gas, occupying a space of about

23,000 cubic feet. If coal gas is used, the size of the balloon would have to be much greater. If hot air is used — as has sometimes been the case — let us assume the temperature of the air inside the envelope such that the density is just half that of the outside air. This would require a temperature probably about 500°. The air

(Photo by Paul Thompson, N.Y.)
AIR BALLOON
Built by some Germans in the backwoods of South Africa

needed would be just a ton, and the balloon would be of about 52,000 cubic feet. It would soon lose its lifting power as the air cooled; and such a balloon would be useful only for short flights.

The 23,000 cubic foot hydrogen balloon, designed to carry a ton, would just answer to sustain the weight. If anchored at sea level, it would neither fall to the ground nor tug upward on its holding-down ropes. In order to ascend, something more is necessary. This "something more" might be some addition to the size and to the amount of hydrogen. Let us assume that we, instead, drop one hundred pounds of our load. Thus relieved of so much ballast, the balloon starts upward, under the net lifting force of one hundred pounds. It is easy to calculate how far it will go. It will not ascend indefinitely, because, as the altitude increases, the pressure (and consequently the density) of the external atmosphere decreases. At about a 2000-foot elevation, this decrease in density will have been sufficient to decrease the buoyant power of the hydrogen to about 1900 pounds, and the balloon will cease to rise, remaining at this level while it moves before the wind.

There are several factors to complicate any calculations. Any expansion of the gas bag — stretching due to an increase in internal pressure — would be one; but the envelope fabrics do not stretch much; there is indeed a very good reason why they must not be allowed to stretch. The pressure in the gas bag is a factor. If there is no stretching of the bag, this pressure will vary directly with the temperature of the gas, and might easily become excessive when the sun shines on the envelope.

A more serious matter is the increased difference between the internal pressure of the gas and the external pressure of the atmosphere at high altitudes. Atmospheric pressure decreases as we ascend. The difference between gas pressure and air pressure thus increases, and it is this difference of pressure which tends to burst the envelope. Suppose the difference of pressure at sea level to have been two-tenths of a pound. For a balloon of twenty feet diameter, this would give a stress on the fabric, per lineal inch, of twenty-four pounds. At an altitude of 2000 feet, the atmospheric pressure would decrease by one pound, the difference of pressures would become one and two-tenths pounds, and the stress on the fabric would be 144 pounds per lineal inch — an absolutely unpermissible strain. There is only one remedy: to allow some of the gas to escape through the safety valve; and this will decrease our altitude.

Ascending and Descending

To ascend, then, we must discard ballast: and we cannot ascend beyond a certain limit on account of the limit of allowable pressure on the envelope fabric. To again descend, we must discharge some of the gas which gives us lifting power. Every change of altitude thus involves a loss either of gas or of ballast. Our vertical field of control may then be represented by a series of oscillations of gradually decreasing magnitude until finally all power to ascend is gone. And even this situation, serious as it is, is made worse by the gradual but steady leakage of gas through the envelope fabric. Here, in a word, is the whole problem of altitude regulation. Air has no surface of equilibrium like water. Some device supplementary to ballast and the safety valve is absolutely necessary for practicable flight in any balloon not staked to the ground.

A writer of romance has equipped his aeronautic heroes with a complete gas-generating plant so that all losses might be made up; and in addition, heating arrangements were provided so that when the gas supply had been partially expended its lifting power could be augmented by warming it so as to decrease its density below even the normal. There might be something to say in favor of this latter device, if used in connection with a collapsible gas envelope.

Methods of mechanically varying the size of the balloon, so as by compressing the gas to cause descent and by giving it more room to increase its lifting power and produce ascent, have been at least suggested. The idea of a vacuum balloon, in which a rigid hollow shell would be exhausted of its contents by a continually working pump, may appear commendable. Such a balloon would have maximum lifting power for its size; but the weight of any rigid shell would be considerable, and the pressure tending to rupture it would be about 100 times that in ordinary gas balloons.

It has been proposed to carry stored gas at high pressure (perhaps in the liquefied condition) as a supplementary method of prolonging the voyage while facilitating vertical movements: but hydrogen gas at a pressure of a ton to the square inch in steel cylinders would give an ultimate lifting power of only about one-tenth the weight of the cylinders which contain it. These cylinders might be regarded as somewhat better than ordinary ballast: but to throw them away, with their gas charge, as ballast, would seem too tragic. Liquefied gas might possibly appear rather more desirable, but would be altogether too expensive.

If a screw propeller can be used on a steamship, a dirigible balloon, or an aeroplane to produce forward motion, there is no reason why it could not also be used to produce upward motion in any balloon; and the propeller with its

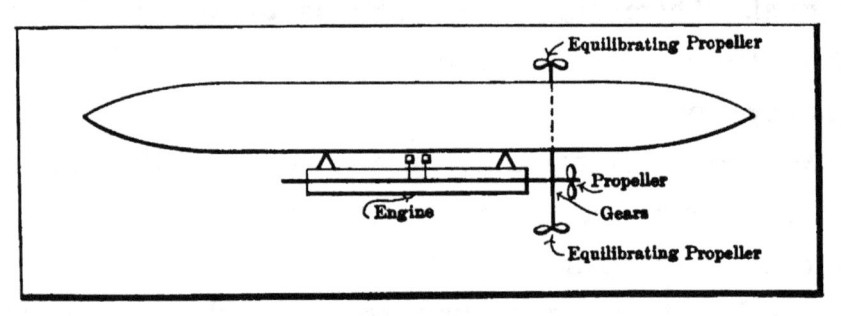

SCREW PROPELLER FOR ALTITUDE CONTROL

operating machinery would be a substitute for twice its equivalent in ballast, since it could produce motion either upward or downward. Weight for weight, however, the propeller and engine give only (in one computed case) about half the lifting power of hydrogen. If we are to use the screw for ascent, we might well use a helicopter, heavier than air, rather than a balloon.

THE BALLONET

The present standard method of improving altitude regulation involves the use of the ballonet, or compartment air bag, inside the main envelope. For stability and effective propulsion, it is important that the balloon preserve its shape, no matter how much gas be allowed to escape. Dirigible balloons are divided into two types, according to the method employed for maintaining the shape. In the Zeppelin type, a rigid internal metal framework supports the gas envelope. This forms a series of seventeen compartments, each isolated from the others. No matter what the pressure of gas, the shape of the balloon is unchanged.

In the more common form of balloon, the internal air ballonet is empty, or nearly so, when the main envelope is full. As gas is vented from the latter, air is pumped into the former. This compresses the remaining gas and thus preserves the normal form of the balloon outline.

But the air ballonet does more than this. It provides

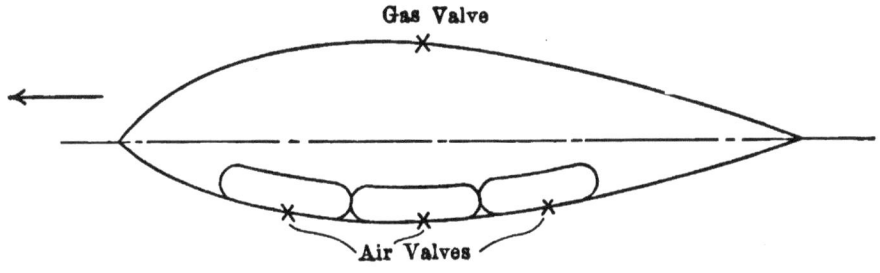

BALLOON WITH BALLONETS

an opportunity for keeping the balloon on a level keel, for by using a number of compartments the air can be circulated from one to another as the case may require, thus altering the distribution of weights. Besides this, if

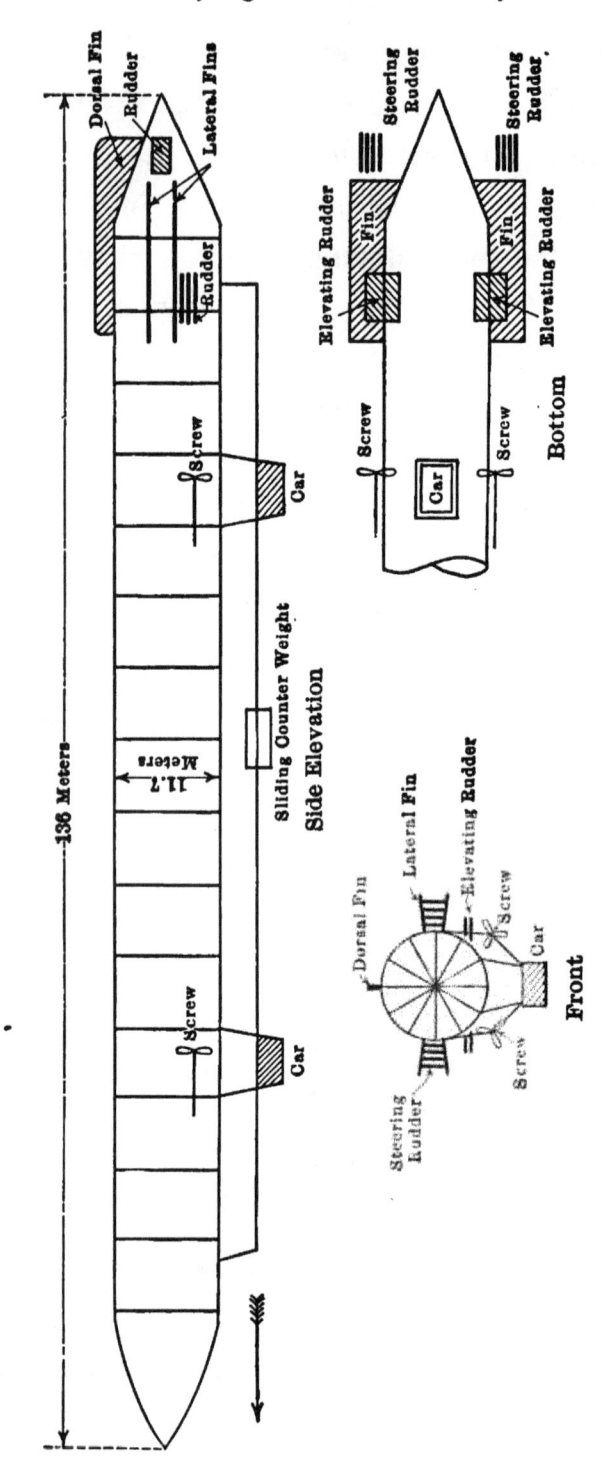

CONSTRUCTION OF THE ZEPPELIN BALLOON

Gas and Ballast

the pressure in the air ballonet be initially somewhat greater than that of the external atmosphere, a considerable ascent may be produced by merely venting this air ballonet. This involves no loss of gas; and when it is again desired to descend, air may be pumped into the ballonet. If any considerable amount of gas should be vented, to produce quick and rapid descent, the pumping of air into the ballonet maintains the shape of the balloon and also facilitates the descent.

The Equilibrator

Suppose a timber block of one square foot area, ten feet long, weighing 380 pounds, to be suspended from the

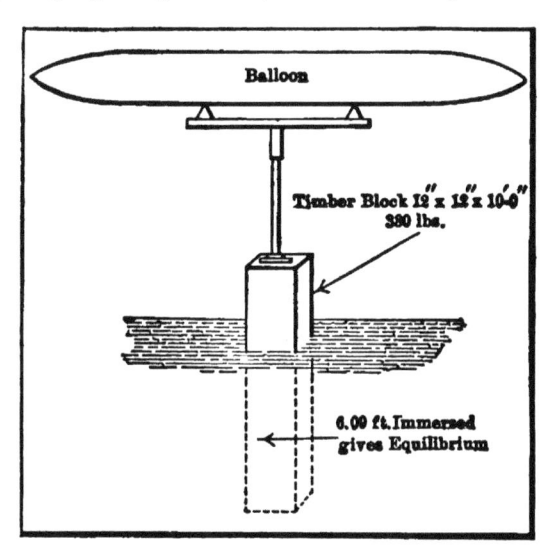

THE EQUILIBRATOR IN NEUTRAL POSITION

balloon in the ocean, and let mechanism be provided by which this block may be raised or lowered at pleasure. When completely immersed in water it exerts an upward pressure (lifting force) of 240 pounds, which may be used to supplement the lifting power of the balloon. If wholly withdrawn from the water, it pulls down the balloon with its weight of 380 pounds. It seems to be equivalent, therefore, to about 620 pounds of ballast. When immersed a little over six feet — the upper four feet being out of the water — it exerts neither lifting nor depressing effect. The amount of either may be perfectly adjusted between the limits stated by varying the immersion.

In the Wellman-Vaniman equilibrator attached to the balloon *America*, which last year carried six men (and a cat) a thousand miles in three days over the Atlantic Ocean, a string of tanks partly filled with fuel was used in place of the timber block. As the tanks were emptied, the degree of control was increased; and this should apparently have given ideal results, equilibration being augmented as the gas supply was lost by leakage: but the unsailorlike disregard of conditions resulting from the strains transferred from a choppy sea to the delicate gas bag led to disaster, and it is doubtful whether this method of control can ever be made practicable. The *America's* trip was largely one of a drifting rather than of a dirigible balloon. The equilibrator could be used only in flights over water in any case: and if we are to look to water for our buoyancy, why not look wholly to water and build a ship instead of a balloon?

DIRIGIBLE BALLOONS AND OTHER KINDS

Shapes

THE cylindrical Zeppelin balloon with approximately conical ends has already been shown (page 68). Those balloons in which the shape is maintained by internal

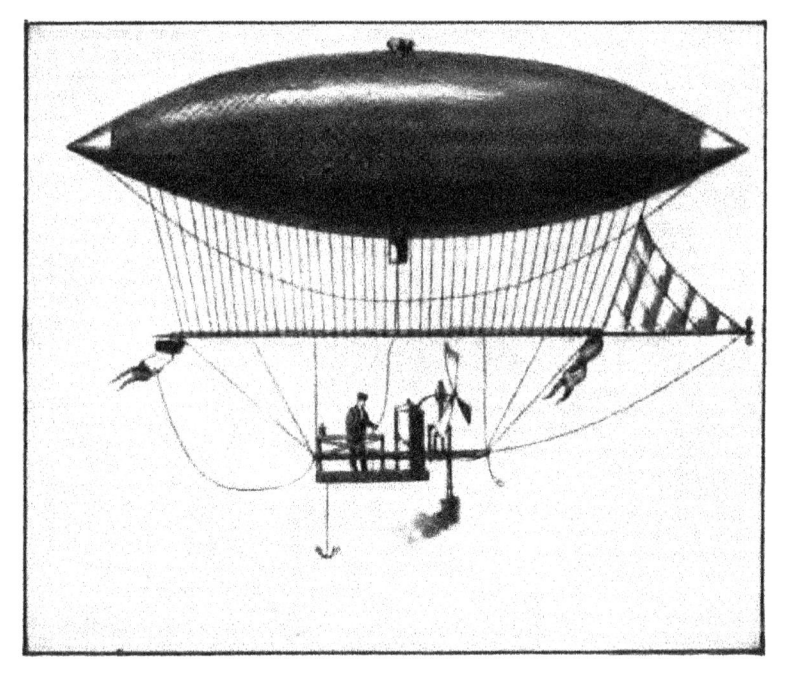

HENRY GIFFARD'S DIRIGIBLE
(The first with steam power)

pressure of air are usually *pisciform*, that is, fish-shaped. Studies have actually been made of the contour lines of various fishes and equivalent symmetrical forms derived, the outline of the balloon being formed by a pair of approximately parabolic curves.

The first flight in a power driven balloon was made by Giffard in 1852. This balloon had an independent speed of about ten feet per second, but was without appliances

DIRIGIBLE OF DUPUY DE LOME
(Man Power)

for steering. A ballonetted balloon of 120,000 cubic feet capacity was directed by man power in 1872: eight men turned a screw thirty feet in diameter which gave a speed of about seven miles per hour. Electric motors and storage batteries were used for dirigible balloons in 1883-'84: in the latter year, Renard and Krebs built the first

Dirigible Balloons and Other Kinds

TISSANDIER BROTHERS' DIRIGIBLE BALLOON
(Electric Motor)

fish-shaped balloon. The first dirigible driven by an internal combustion motor was used by Santos-Dumont in 1901.

DIMENSIONS

The displacements of present dirigibles vary from 20,000 cubic feet (in the United States Signal Corps airship) up to 460,000 cubic feet (in the Zeppelin). The former balloon has a carrying capacity only about equivalent to that of a Wright biplane. While anchored or drifting balloons are usually spherical, all dirigibles are elongated, with a length of from four to eleven diameters. The Zeppelin represents an extreme elongation, the length being 450 feet and the diameter forty-two feet. At the other extreme, some of the English military dirigibles are thirty-one feet in diameter and only 112 feet long. Ballonet capacities may run up to one-fifth the gas volume. All present dirigibles have gasoline engines driving propellers from

THE BALDWIN
Dirigible of the United States Signal Corps

eight to twenty feet in diameter. The larger propellers are connected with the motors by gearing, and make from 250 to 700 turns per minute. The smaller propellers are direct connected and make about 1200 revolutions. Speeds are usually from fifteen to thirty miles per hour.

The present-day elongated shape is the result of the effort

Dirigible Balloons and Other Kinds 75

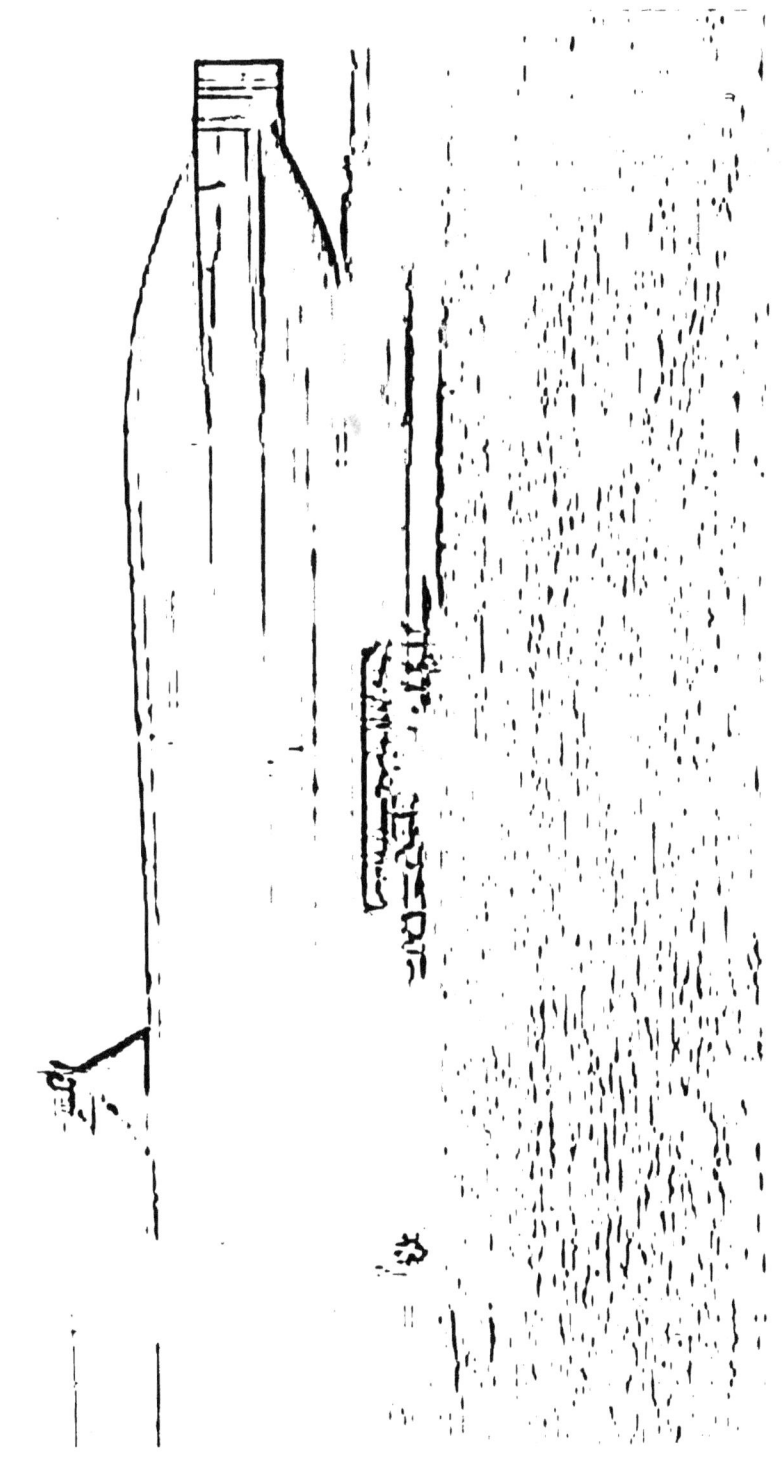

THE ZEPPELIN ENTERING ITS HANGAR ON LAKE CONSTANCE

to decrease the proportion of propulsion resistance due to the pressure of the air against the head of the balloon. This has led also to the pointed ends now universal; and to avoid eddy resistance about the rear it is just as important to point the stern as the bow. As far as head end resistance alone is concerned, the longer the balloon the better: but the friction of the air along the side of the envelope also produces resistance, so that the balloon must not be too much elongated. Excessive elongation also produces structural weakness. From the standpoint of stress on the fabric of the envelope, the greatest strain is that which tends to break the material along a longitudinal line, and this is true no matter what the length, as long as the seams are equally strong in both directions and the load is so suspended as not to produce excessive bending strain on the whole balloon. In the *Patrie* (page 77), some distortion due to loading is apparent. The stress per lineal inch of fabric is obtained by multiplying the net pressure by half the diameter of the envelope (in inches).

Ample steering power (provided by vertical planes, as in heavier-than-air machines) is absolutely necessary in dirigibles: else the head could not be held up to the wind and the propelling machinery would become ineffective.

Fabrics

The material for the envelope and ballonets should be light, strong, unaffected by moisture or the atmosphere, non-cracking, non-stretching, and not acted upon by varia-

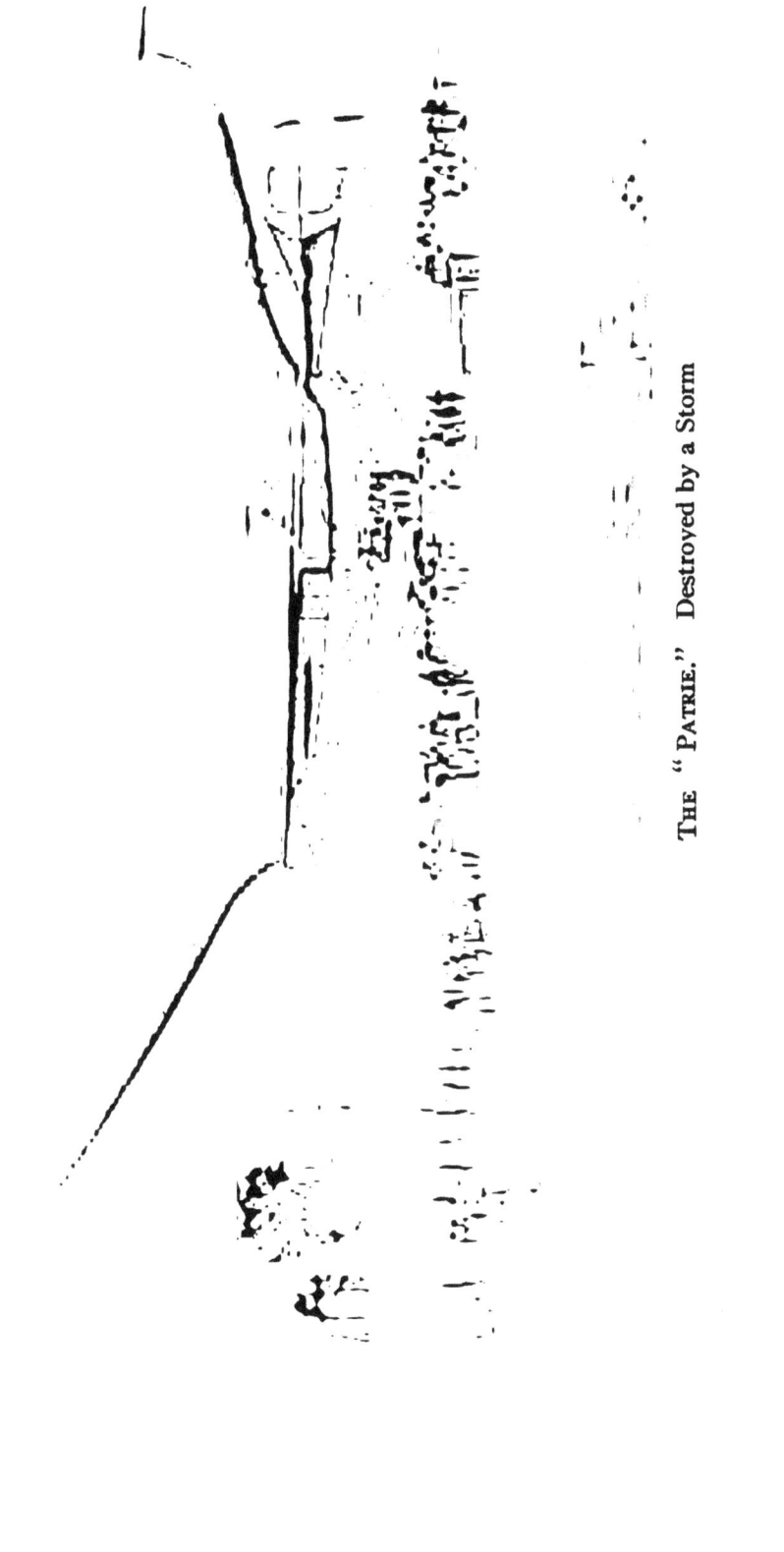

The "Patrie." Destroyed by a Storm

tions in temperature. The same specifications apply to the material for the wings of an aeroplane. In addition, for use in dirigible balloons, fabrics must be impermeable, resistent to chemical action of the gas, and not subject to spontaneous combustion. The materials used are vulcanized silk, gold beater's skin, Japanese silk and rubber, and cotton and rubber compositions. In many French balloons, a middle layer of rubber has layers of cotton on each side, the whole thickness being the two hundred and fiftieth part of an inch. In the *Patrie*, this was supplemented by an outside non-heat-absorbent layer of lead chromate and an inside coating of rubber, all rubber being vulcanized. The inner rubber layer was intended to protect the fabric against the destructive action of impurities in the gas.

Fabrics are obtainable in various colors, painted, varnished, or wholly uncoated. The rubber and cotton mixtures are regularly woven in France and Germany for aeroplanes and balloons. The cars and machinery are frequently shielded by a fabricated wall. Weights of envelope materials range from one twenty-third to one-fourteenth pound per square foot, and breaking stresses from twenty-eight to one hundred and thirty pounds. Pressures (net) in the main envelope are from three-fifths to one and a quarter *ounces* per square inch, those in the ballonets being somewhat less. The *Patrie* of 1907 had an envelope guaranteed not to allow the leakage of more than half a cubic inch of hydrogen per square foot of surface per twenty-four hours.

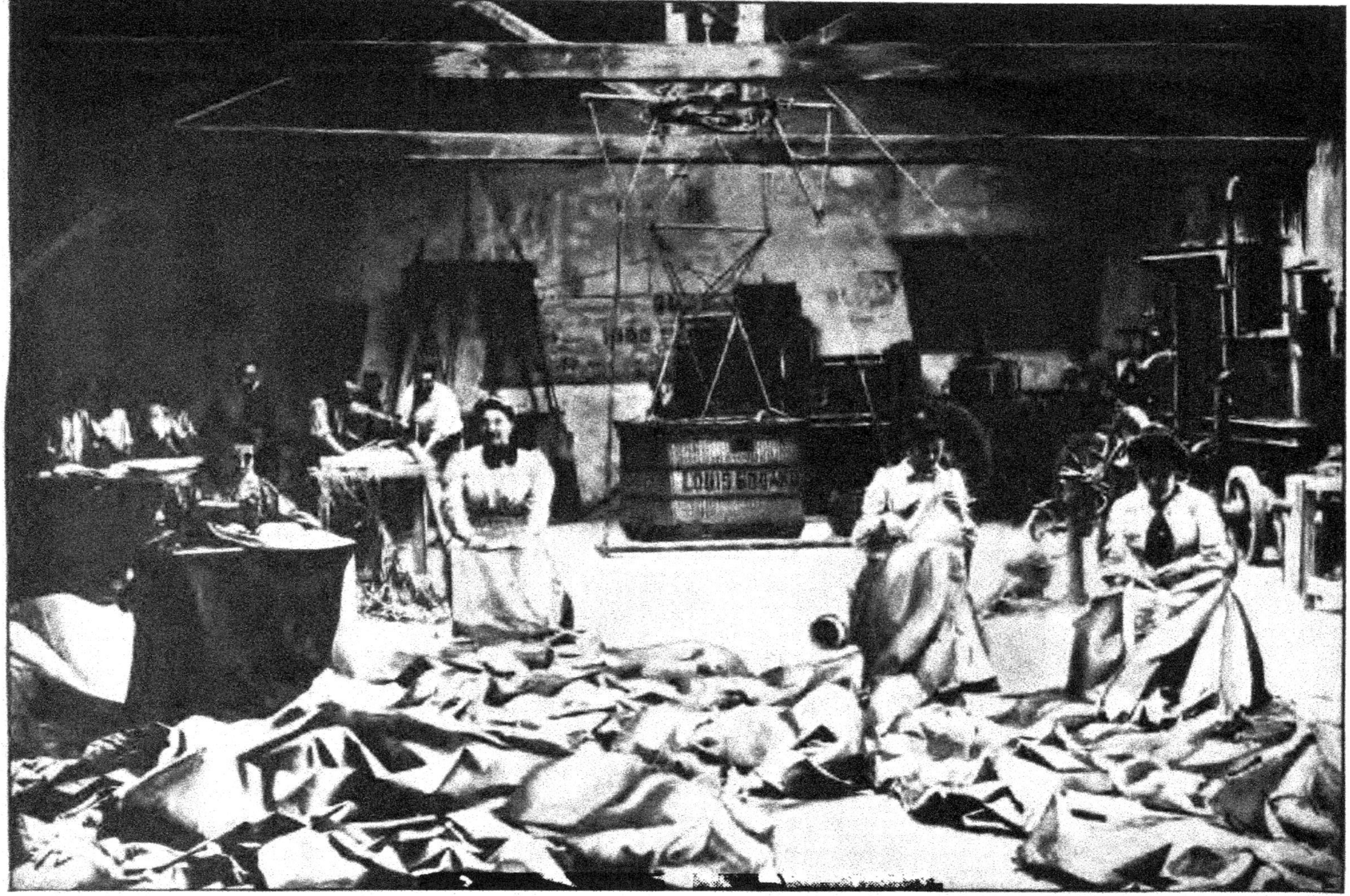

MANUFACTURING THE ENVELOPE OF A BALLOON.

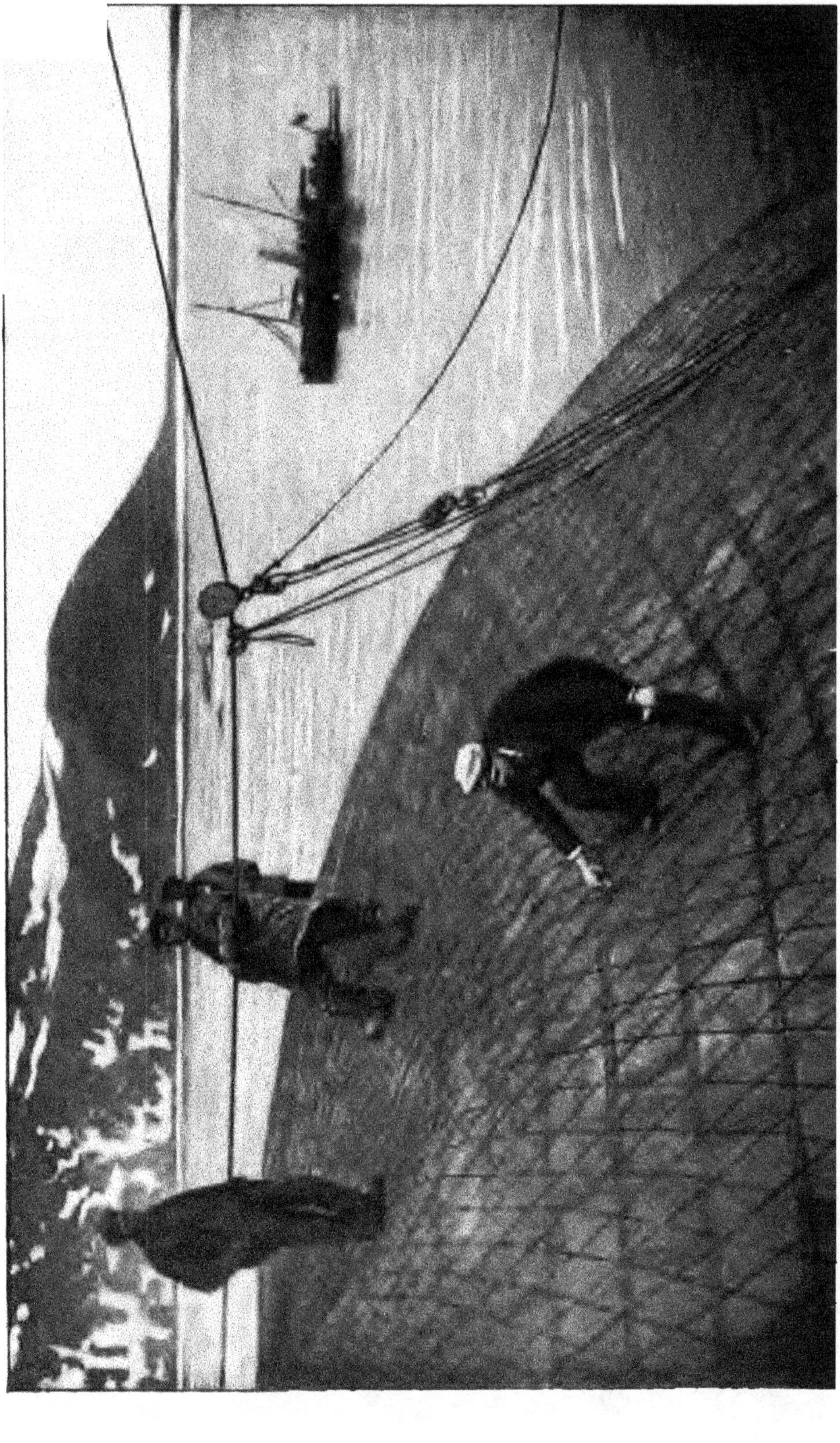

The best method of cutting the fabric is to arrange for building up the envelope by a series of strips about the circumference, the seams being at the bottom. The two warps of the cloth should cross at an angle so as to localize a rip or tear. Bands of cloth are usually pasted over the seams, inside and out, with a rubber solution; this is to prevent leakage at the stitches.

FRAMING

In the *Zeppelin*, the rigid aluminum frame is braced every forty-five feet by transverse diametral rods which make the cross-sections resemble a bicycle wheel (page 68). This cross-section is not circular, but sixteen-sided. The pressure is resisted by the framework itself, the envelope being required to be impervious only. The seventeen compartments are separated by partitions of sheet aluminum. There is a system of complete longitudinal bracing between these partitions. Under the main framework, the cars and machinery are carried by a truss about six feet deep which runs the entire length. The cars are boat-shaped, twenty feet long and six feet wide, three and one-half feet high, enclosed in aluminum sheathing. These cars, placed about one hundred feet from the ends, are for the operating force and machinery. The third car, carrying passengers, is built into the keel.

In non-rigid balloons like the *Patrie*, the connecting frame must be carefully attached to the envelope. In this particular machine, cloth flaps were sewed to the bag, and

WRECK OF THE "ZEPPELIN"

nickel steel tubes then laced in the flaps. With these tubes as a base, a light framework of tubes and wires, covered with a laced-on waterproof cloth, was built up for supporting the load. Braces ran between the various stabilizing and controlling surfaces and the gas bag; these were for the most part very fine wire cables. The weight of the car was concentrated on about seventy feet of the total length of 200 feet. This accounts for the deformation of the envelope shown in the illustration (page 77). The frame and car of this balloon were readily dismantled for transportation.

In some of the English dirigibles the cars were suspended by network passing over the top of the balloon.

KEEPING THE KEEL HORIZONTAL

In the *Zeppelin*, a sliding weight could be moved along the keel so as to cause the center of gravity to coincide with the center of upward pressure in spite of variations in weight and position of gas, fuel, and ballast. In the German balloon *Parseval*, the car itself was movable on a longitudinal suspending cable which carried supporting sheaves. This balloon has figured in recent press notices. It was somewhat damaged by a collision with its shed in March: the sixteen passengers escaped unharmed. A few days later, emergency deflation by the rip-strip was made necessary during a severe storm. In the ordinary non-rigid balloon, the pumping of air between the ballonets aids in controlling longitudinal equilibrium. The pump may be

CAR OF THE ZEPPELIN
(From the *Transactions* of the American Society of Mechanical Engineers)

arranged for either hand or motor operation: that in the *Clément-Bayard* had a capacity of 1800 liters per minute against the pressure of a little over three-fifths of an ounce. The *Parseval* has two ballonets. Into the rear of these air is pumped at starting. This raises the bow and facilitates ascent on the principle of the inclined surface of an aeroplane. After some elevation is attained, the forward ballonet is also filled.

STABILITY

Besides proper distribution of the loads, correct vertical location of the propeller is important if the balloon is to travel on a level keel. In some early balloons, two envelopes side by side had the propeller at the height of the axes of the gas bags and midway between them. The modern forms carry the car, motor, and propeller below the balloon proper. The air resistance is mostly that of the bow of the envelope: but there is some resistance due to the car, and the propeller shaft should properly be at the equivalent center of all resistance, which will be between car and axis of gas bag and nearer the latter than the former. With a single envelope and propeller, this arrangement is impracticable. By using four (or even two) propellers, as in the *Zeppelin* machine (page 68), it can be accomplished. If only one propeller is employed, horizontal rudder planes must be disposed at such angles and in such positions as to compensate for the improper position of the tractive force. Even on the *Zeppelin*, such planes were employed with advantage (pages 66 and 73).

Perfect stability also involves freedom from rolling. This is usually inherent in a balloon, because the center of mass is well below the center of buoyancy: but in machines of the non-rigid type the absence of a ballonet might lead

STERN VIEW OF THE ZEPPELIN

to both rolling and pitching when the gas was partially exhausted.

What is called "route stability" describes the condition of straight flight. The balloon must point directly in its (independent) course. This involves the use of a steering rudder, and, in addition, of fixed vertical planes, which, on the principle of the vertical partitions of Voisin, probably give some automatic steadiness to the course. To avoid the difficulty or impossibility of holding the head up to the wind at high speeds, an *empennage* or feathering tail

THE "CLÉMENT-BAYARD"

is a feature of all present balloons. The empennage of the *Patrie* (page 77) consisted of pairs of vertical and horizontal planes at the extreme stern. In the *France*, thirty-two feet in maximum diameter and nearly 200 feet long, empennage planes aggregating about 400 square feet were placed somewhat forward of the stern. In the *Clément-*

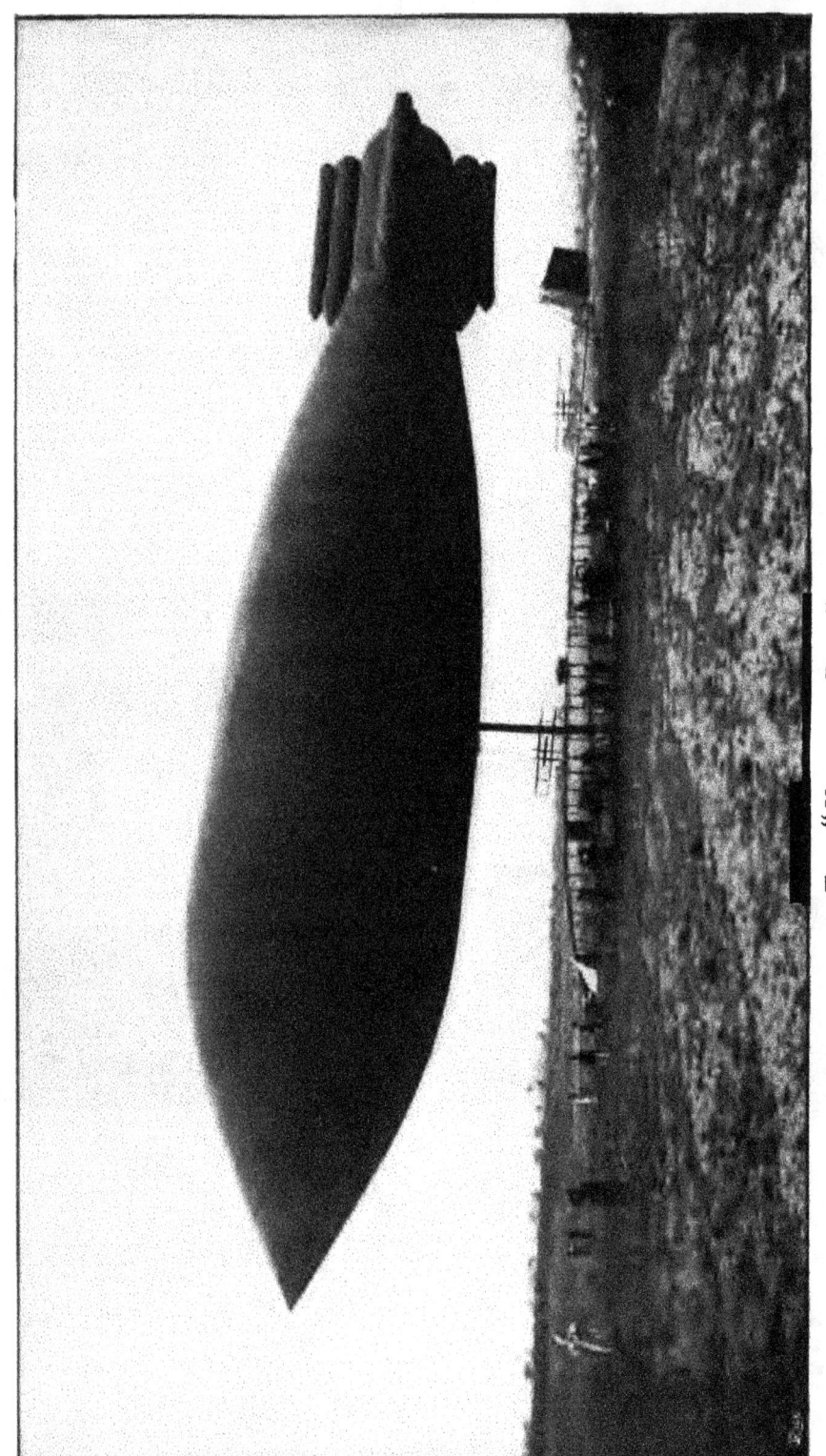

The "Ville de Paris"

Bayard, the empennage consisted of cylindro-conical ballonets projecting aft from the stern. A rather peculiar grouping of such ballonets was used about the prolonged stern of the *Ville de Paris.*

RUDDERS AND PLANES

The dirigible has thus several air-resisting or gliding surfaces. The approximately "horizontal" (actually some-

CAR OF THE "LIBERTÉ"

what inclined) planes permit of considerable ascent and descent by the expenditure of power rather than gas, and thus somewhat influence the problem of altitude control. Each of the four sets of horizontal rudder planes on the *Zeppelin*, for example, has, at thirty-five miles per hour, with an inclination equal to one-sixth a right angle, a lifting power of nearly a ton; about equal to that of all of the gas in one of the sixteen compartments.

Movable rudders may be either hand or motor-operated. The double vertical steering rudder of the *Ville de Paris* had an area of 150 square feet. The horizontally pivoted rudders for vertical direction had an area of 130 square feet.

Arrangement and Accessories

The motor in the *Ville de Paris* was at the front of the car, the operator behind it. This car had the excessive weight of nearly 700 pounds. The *Patrie* employed a non-combustible shield over the motor, for the protection of the envelope: its steering wheel was in front and the motor about in the middle of the car. The gasoline tank was under the car, compressed air being used to force the fuel up to the motor, which discharged its exhaust downward at the rear through a spark arrester. Motors have battery and magneto ignition and decompression cocks, and are often carried on a spring-supported chassis. The interesting *Parseval* propeller has four cloth blades which hang limp when not revolving. When the motor is running, these blades, which are weighted with lead at the proper points, assume the desired form.

Balloons usually carry guide ropes at head and stern, the aggregate weight of which may easily exceed a hundred pounds. In descending, the bow rope is first made fast, and the airship then stands with its head to the wind, to be hauled in by the stern rope. For the large French military balloons, this requires a force of about thirty men. The *Zeppelin* descends in water, being lowered until the cars float, when it is docked like a ship (see page 84). Landing skids are sometimes used, as with aeroplanes.

The balloon must have escape valves in the main envelope and ballonets. In addition it has a "rip-strip" at the bottom by which a large cut can be made and the gas quickly vented for the purpose of an emergency descent. Common equipment includes a siren, megaphone, anchor pins, fire extinguisher, acetylene search light, telephotographic apparatus, registering and indicating gages and other instruments, anemometer, possibly carrier pigeons; besides fuel, oil and water for the motor, and the necessary supplies for the crew. The glycerine floated compass of Moisant must now also be included if we are to contemplate genuine navigation without constant recourse to landmarks.

Amateur Dirigibles

The French Zodiac types of "aerial runabout" displace 700 cubic meters, carrying one passenger with coal gas or two passengers with a mixture of coal gas and hydrogen. The motor is four-cylinder, sixteen horse-power, water-cooled. The stern screw, of seven feet diameter, makes 600 turns per minute, giving an independent speed of nineteen miles per hour. The machine can remain aloft three hours with 165 pounds of supplies. It costs $5000. Hydrogen costs not far from a cent per cubic foot (twenty

92   *Flying Machines Today*

THE ZODIAC No. 2
May be deflated and easily transported

cents per cubic meter) so that the question of gas leakage may be at least as important as the tire question with automobiles.

THE FORT OMAHA PLANT

The Signal Corps post at Fort Omaha has a plant comprising a steel balloon house of size sufficient to house one of the largest dirigibles built, an electrolytic plant for generating hydrogen gas, having a capacity of 3000 cubic feet per hour, a 50,000 cubic foot gas storage tank, and the compressing and carrying equipment involved in preparing gas for shipment at high pressure in steel cylinders.

UNITED STATES SIGNAL CORPS BALLOON PLANT AT FORT OMAHA, NEB.
(From the *Transactions* of the American Society of Mechanical Engineers)

Balloon Progress

The first aerial buoy of Montgolfier brothers, in 1783, led to the suggestion of Meussier that two envelopes be used; the inner of an impervious material to prevent gas leakage, and the outer for strength. There was perhaps

THE "CAROLINE" OF ROBERT BROTHERS, 1784
The ascent terminated tragically

a foreshadowing of the Zeppelin idea. Captive and drifting balloons were used during the wars of the French Revolution: they became a part of standard equipment in our own War of Secession and in the Franco-Prussian conflict. The years 1906 to 1908 recorded rapid progress in the development of the dirigible: the record-breaking *Zeppelin* trip was in 1909 and Wellman's *America* exploit

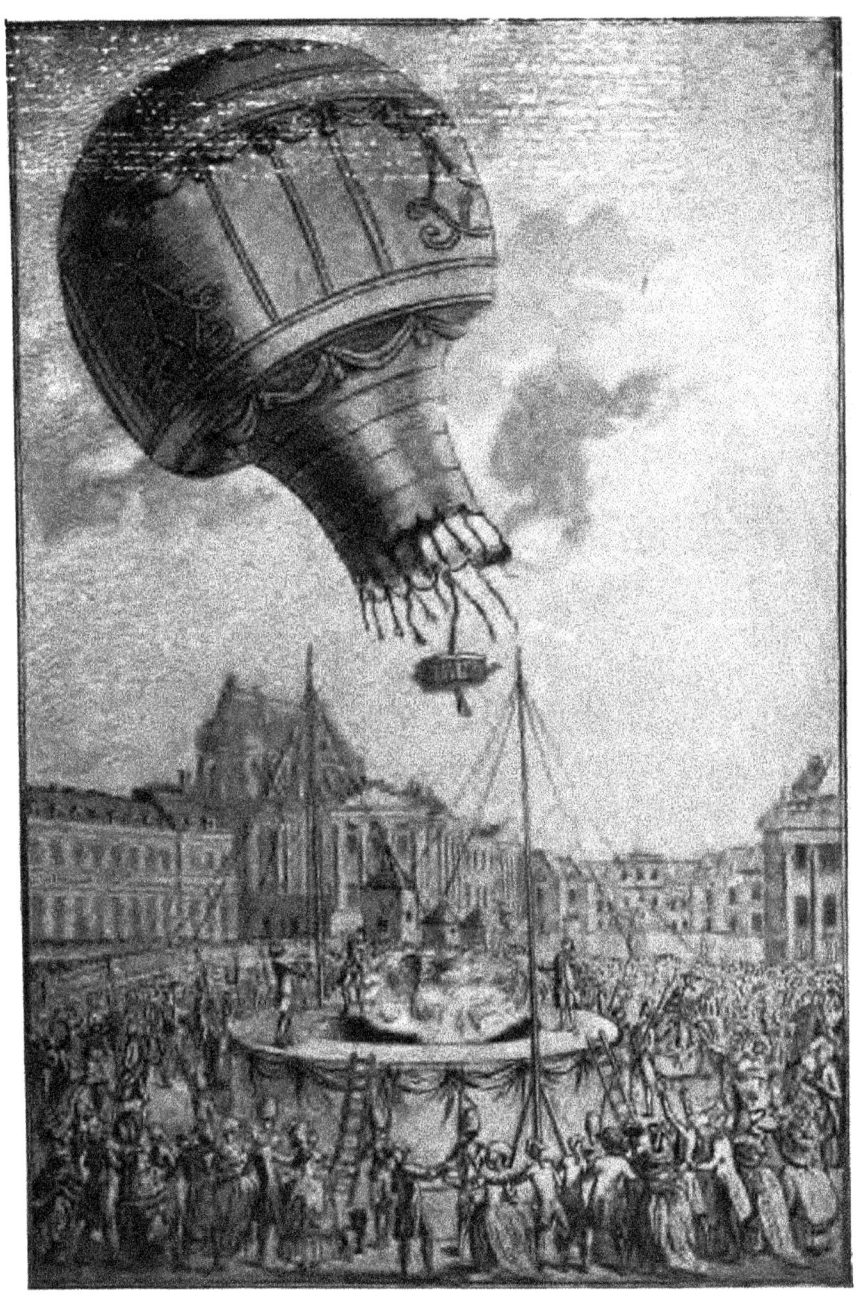

THE ASCENT AT VERSAILLES, 1783
The first balloon carrying living beings in the air

in October, 1910. Unfortunately, dirigibles have had a a bad record for stanchness: the *Patrie, Republique, Zeppelin* (*I* and *II*), *Deutschland, Clément-Bayard* — all have gone to that bourne whence no balloon returns.

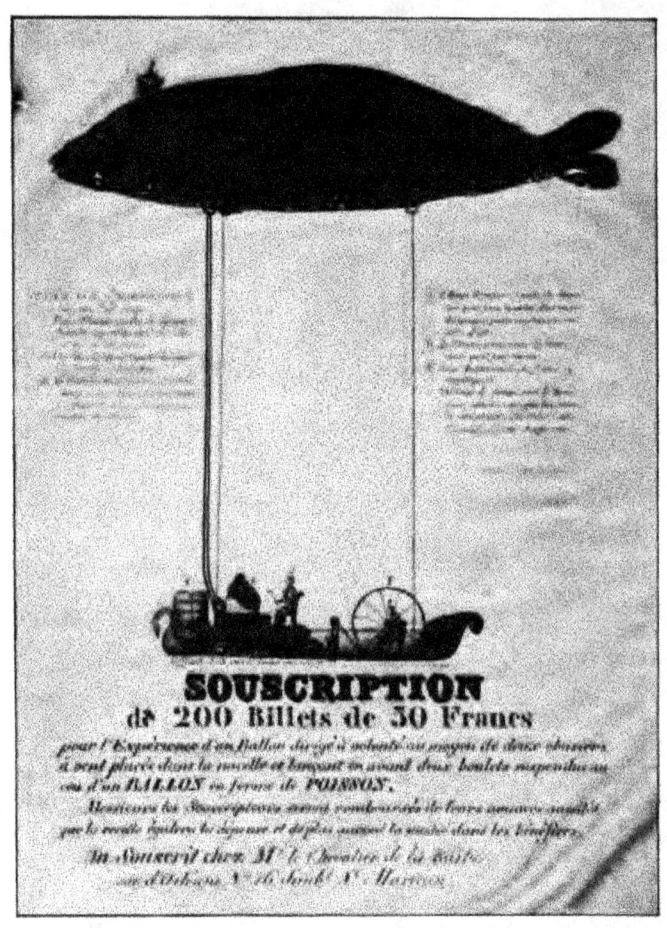

Investors were lacking to bring about the realization of this project

It is gratifying to record that Count Zeppelin's latest machine, the *Deutschland II*, is now in operation. During the present month (April, 1911), flights have been made covering 90 miles and upward at speeds exceeding 20 miles

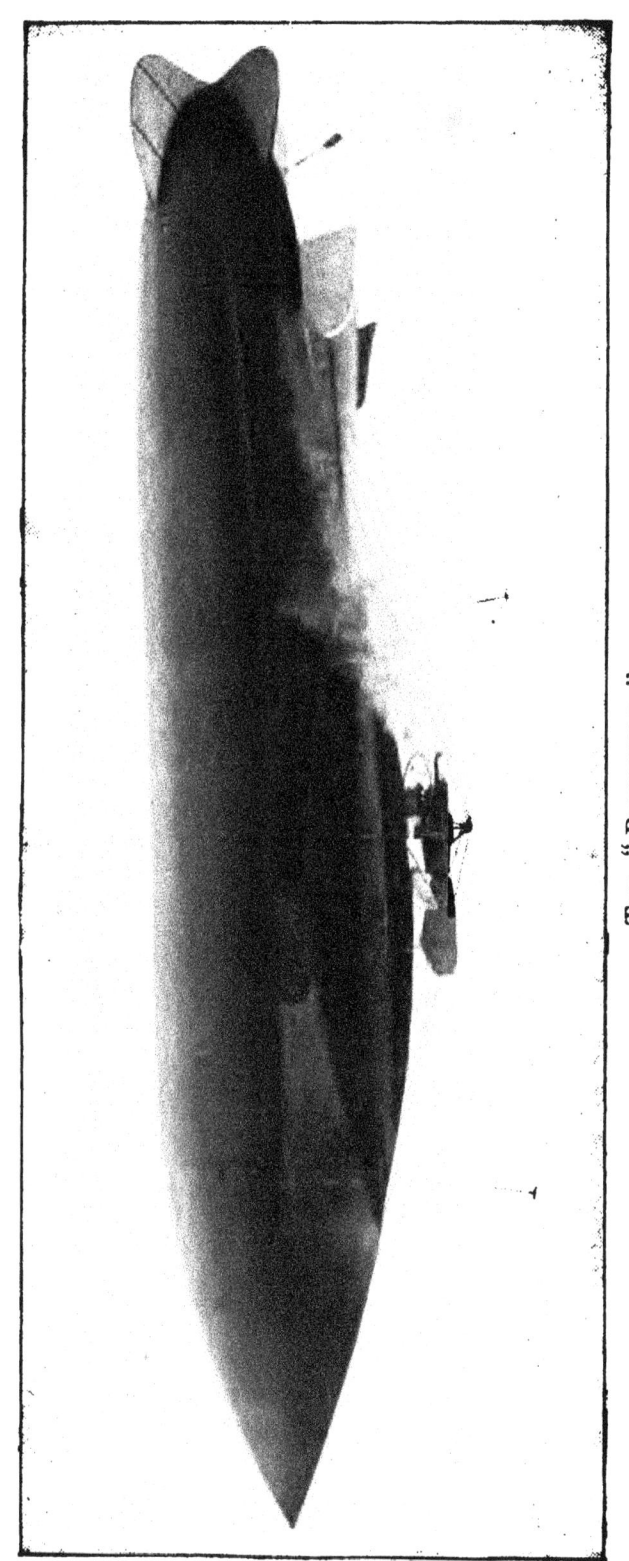

The "Republique"

per hour with the wind unfavorable. This balloon is intended for use as a passenger excursion vehicle during the coming summer, under contract with the municipality of Düsseldorf.

At the present moment, Neale, in England, is reported to be building a dirigible for a speed of a hundred miles per hour. The Siemens-Schuckart non-rigid machine, nearly 400 feet long and of 500 horse-power, is being tried out at Berlin: it is said to carry fifty passengers.* Fabrice, of Munich, is experimenting with the *Inchard*, with a view to crossing the Atlantic at an early date. Mr. Vaniman, partner of Wellman on the *America* expedition, is planning a new dirigible which it is proposed to fly across the ocean before July 4. The engine, according to press reports, will develop 200 horse-power, and the envelope will be more elongated than that of the *America*. And meanwhile a Chicago despatch describes a projected fifty-passenger machine, to have a gross lifting power of twenty-five tons!

Germany has a slight lead in number of dirigible balloons — sixteen in commission and ten building. France follows closely with fourteen active and eleven authorized. This accounts for two-thirds of all the dirigible balloons in the world. Great Britain, Italy, and Russia rank in the order named. The United States has one balloon of the smallest size. Spain has, or had, one dirigible. As to

---

* According to press reports, temporary water ballast will be taken on during the daytime, to offset the ascensional effect of the hot sun on the envelope.

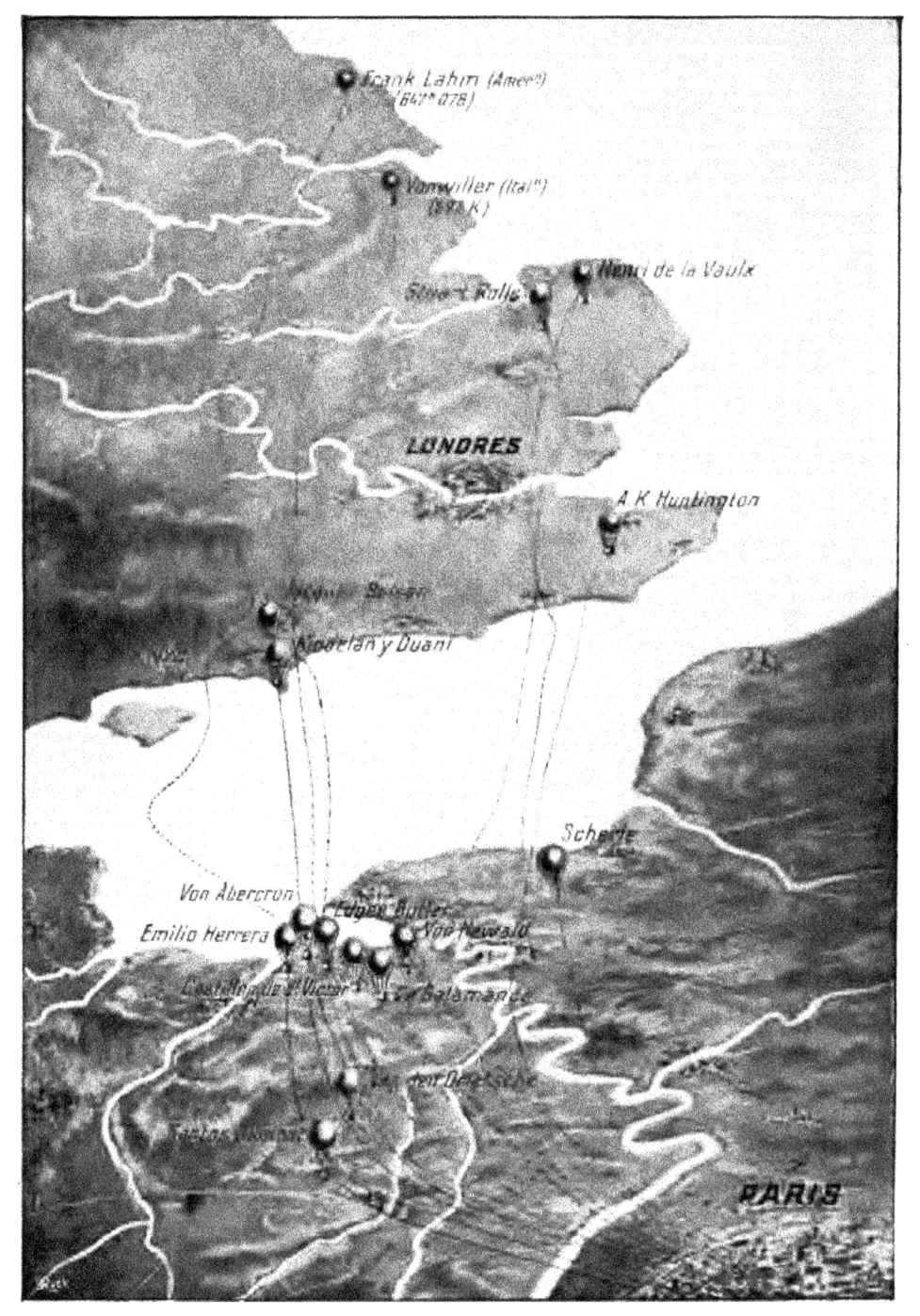

THE FIRST FLIGHT FOR THE GORDON-BENNET CUP.
Won by Lieut. Frank P. Lahm, U.S.A., 1906. Figures on the map denote distances in kilometers. The cup has been offered annually by Mr. James Gordon-Bennet for international competition under such conditions as may be prescribed by the International Aeronautic Federation.

aeroplanes, however, the United States and England rank equally, having each about one-fourth as many machines as France (which seems, therefore, to maintain a "four-power standard"). Germany, Russia, and Italy follow, in order, the United States. These figures include all machines, whether privately or nationally owned. Until lately, our own government operated but one aeroplane. A recent appropriation by Congress of $125,000 has led to arrangements for the purchase of a few additional biplanes of the Wright and Curtiss types; and a training school for army officers has been regularly conducted at San Diego, Cal., during the past winter. The Curtiss machine to be purchased is said to carry 700 pounds of dead weight with a sail area of 500 square feet. It is completely demountable and equipped with pontoons.

THE QUESTION OF POWER

IN the year 1810, a steam engine weighed something over a ton to the horse-power. This was reduced to about 200 pounds in 1880. The steam-driven dirigible balloon of Giffard, in 1852, carried a complete power plant weighing a little over 100 pounds per horse-power; about the weight of a modern locomotive. The unsuccessful Maxim flying machine of 1894 brought this weight down to less than 20 pounds. The gasoline engine on the original Wright machines weighed about 5 pounds to the horse-power; those on some recent French machines not far from 2 pounds.

Pig iron is worth perhaps a cent a pound. An ordinary steam or gas engine may cost eight cents a pound; a steam turbine, perhaps forty cents. A high grade automobile or a piano may sell for a dollar a pound; the Gnome aeroplane motor is priced at about twenty dollars a pound. This is considerably more than the price of silver. The motor and accessories account for from two-thirds to nine-tenths of the total cost of an aeroplane.

A man weighing 150 pounds can develop at the outside about one-eighth of a horse-power. It would require 1200 pounds of man to exert one horse-power. Considered as an engine, then, a man is (weight for weight) only one six-hundredth as effective as a Gnome motor. In the original Wright aeroplane, a weight of half a ton was sustained at the expenditure of about twenty-five horse-power. The motor weight was about one-eighth of the total weight. If traction had been produced by man-power,

THE GNOME MOTOR
(Aeromotion Company of America)

30,000 pounds of man would have been necessary: thirty times the whole weight supported.

Under the most favorable conditions, to support his own weight of 150 pounds (at very high gliding velocity and a slight angle of inclination, disregarding the weight of sails necessary), a man would need to have the strength of about fifteen men. No such thing as an aerial bicycle, therefore, appears possible. The man can not emulate the bird.

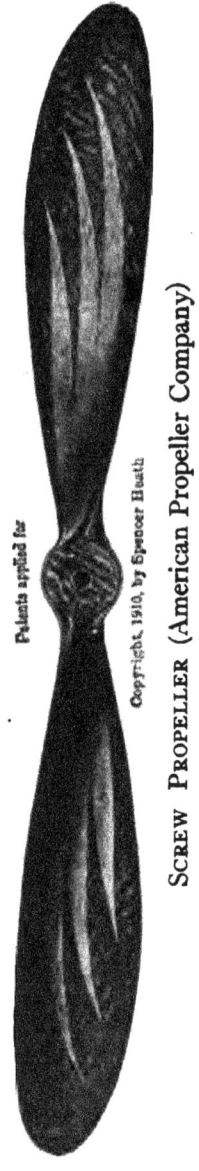

Screw Propeller (American Propeller Company)

The power plant of an air craft includes motor, water and water tank, radiator and piping, shaft and bearings, propeller, controlling wheels and levers, carbureter, fuel, lubricating oil and tanks therefor. Some of the weight may eventually be eliminated by employing a two-cycle motor (which gives more power for its size) or by using rotary air-cooled cylinders. Propellers are made light by employing wood or skeleton construction. One eight-foot screw of

ONE OF THE MOTORS OF THE ZEPPELIN

white oak and spruce, weighing from twelve to sixteen pounds, is claimed to give over 400 pounds of propelling force at a thousand turns per minute.

The cut shows the action of the so-called "four-cycle" motor. Four strokes are required to produce an impulse on the piston and return the parts to their original posi-

The Question of Power 105

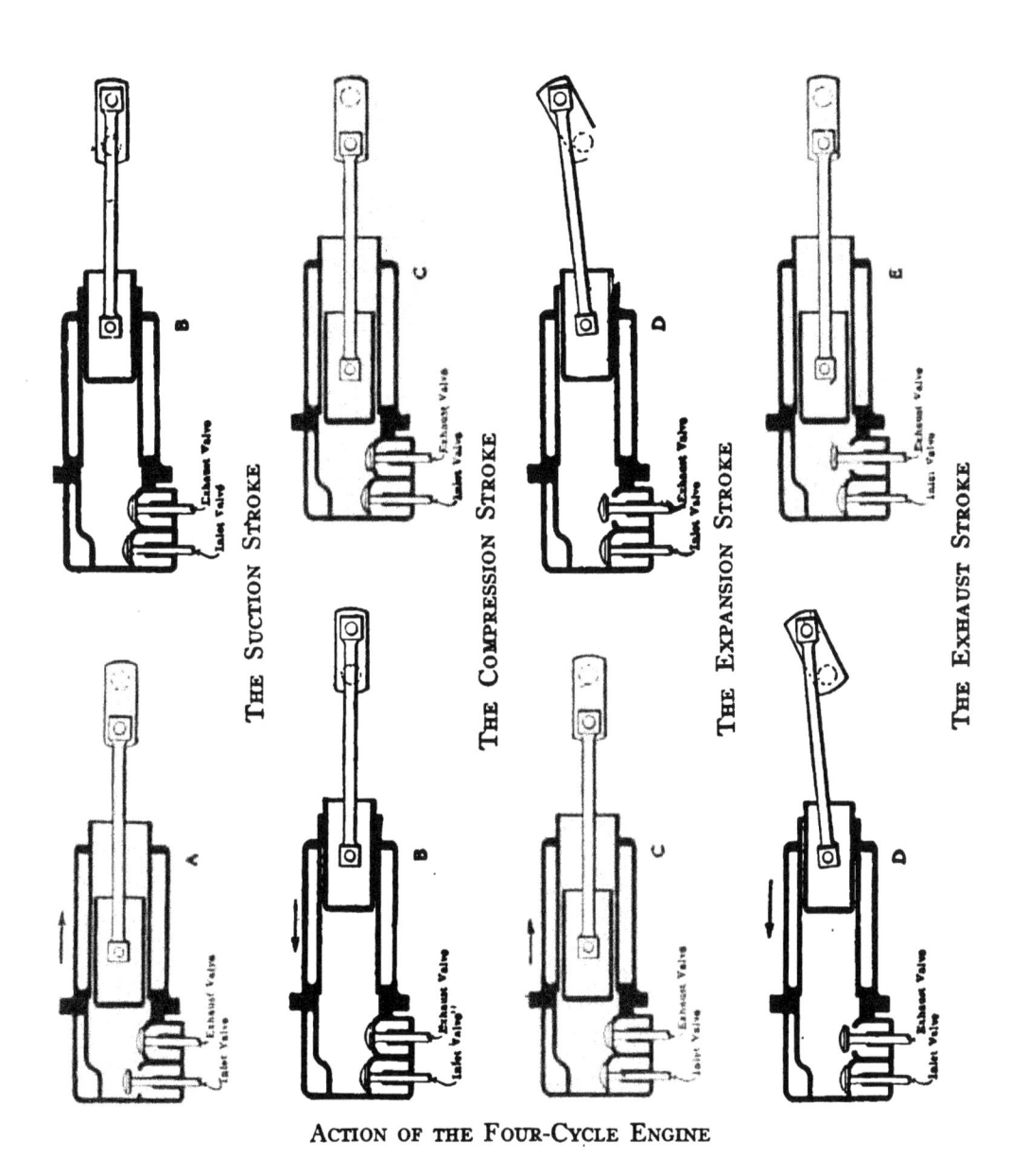

THE SUCTION STROKE   THE COMPRESSION STROKE   THE EXPANSION STROKE   THE EXHAUST STROKE

ACTION OF THE FOUR-CYCLE ENGINE

tions. On the first, or suction stroke, the combustible mixture is drawn into the cylinder, the inlet valve being open and the outlet valve closed. On the second stroke, both valves are closed and the mixture is highly compressed. At about the end of this stroke, a spark ignites the charge, a still greater pressure is produced in consequence, and the energy of the gas now forces the piston outward on its third or "working" stroke, the valves remaining closed. Finally, the outlet valve is opened and a fourth stroke sweeps the burnt gas out of the cylinder.

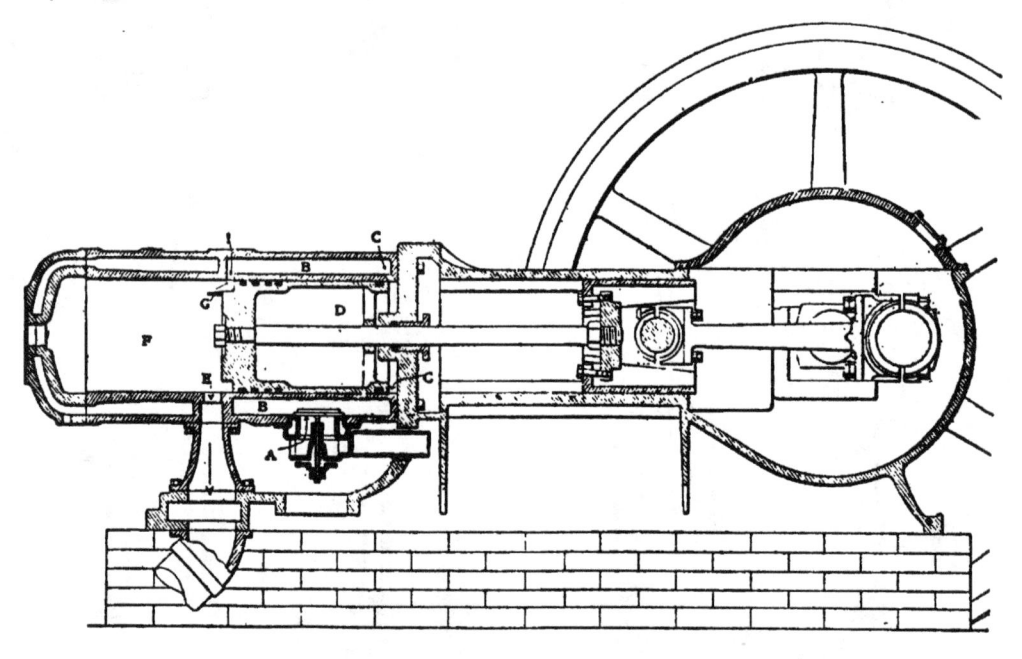

In the "two-cycle" engine, the piston first moves to the left, compressing a charge already present in the cylinder at F, and meanwhile drawing a fresh supply through the valve *A* and passages *C* to the space *D*. On the return stroke, the exploded gas in F expands, doing its work, while that in D is slightly compressed, the valve A being now closed. When the piston, moving toward the right, opens the passage E, the burnt gas rushes out. A little later, when the passage I is exposed, the fresh compressed gas in D rushes through C, B, and I to F. The operation may now be repeated. Only two strokes have been necessary. The cylinder develops power twice as rapidly as before: but at the cost of some waste of gas, since the inlet (I) and outlet (E) passages are for a brief interval *both open at once:* a condition not altogether remedied by the use of a deflector at G. A two-cycle cylinder should give nearly twice the power of a four-cycle cylinder of the same size, and the two-cycle engine should weigh less, per horse-power; but it requires from 10 to 30% more fuel, and fuel also counts in the total weight.

The high temperatures in the cylinder would soon make the cast-iron walls red-hot, unless the latter were artificially cooled. The usual method of cooling is to make the walls hollow and circulate water through them. This involves a pump, a quantity of water, and a "radiator" (cooling machine) so that the water can be used over and over again. To cool by air blowing over the surface of the cylinder is relatively ineffective: but has been made possible in automobiles by building fins on the cylinders so as to increase the amount of cooling surface. When the motors are worked at high capacity, or when two-cycle motors are used, the heat is generated so rapidly that this method of cooling is regarded as inapplicable. By rapidly rotating the cylinders themselves through the air, as in motors like the Gnome, air cooling is made sufficiently adequate, but the expenditure of power in producing this rotation has perhaps not been sufficiently regarded.

MOTOR AND PROPELLER
(Detroit Aeronautic Construction Co.)

Possible progress in weight economy is destined to be limited by the necessity for reserve motor equipment.

The engine used is usually the four-cycle, single-acting, four-cylinder gasoline motor of the automobile, designed for great lightness. The power from each cylinder of such a motor is approximately that obtained by dividing the square of the diameter in inches by the figure 2½. Thus a five-inch cylinder should give ten horse-power — at normal piston speed. On account of friction losses and the wastefulness of a screw propeller, not more than half this power is actually available for propulsion.

The whole power plant of the *Clément-Bayard* weighed about eleven pounds to the horse-power. This balloon was 184 feet long and 35 feet in maximum diameter, displacing about 100,000 cubic feet. It carried six passengers, about seventy gallons of fuel, four gallons of lubricating oil, fifteen gallons of water, 600 pounds of ballast, and 130 pounds of ropes. The motor developed 100 horse-power at a thousand revolutions per minute. About eight gallons of fuel and one gallon of oil were consumed per hour when running at the full independent speed of thirty-seven miles per hour.

The Wellman balloon *America* is said to have consumed half a ton of gasoline per twenty-four hours: an eight days' supply was carried. The gas leakage in this balloon was estimated to have been equivalent to a loss of 500 pounds of lifting power per day.

The largest of dirigibles, the *Zeppelin*, had two motors of 170 horse-power each. It made, in 1909, a trip of over 800 miles in thirty-eight hours.

The engine of the original Voisin cellular biplanes was an eight-cylinder Antoinette of fifty horse-power, set near

Two-Cylinder Opposed Engine.
(From *Aircraft*)

Four-Cylinder Vertical Engine
(The Dean Manufacturing Co.)

the rear edge of the lower of the main planes. The Wright motors are placed near the front edge. A twenty-five horse power motor at 1400 revolutions propelled the Fort Myer machine, which was built to carry two passengers, with fuel for a 125 mile flight: the total weight of the whole flying apparatus being about half a ton.

The eight-cylinder Antoinette motor on a Farman biplane, weighing 175 pounds, developed thirty-eight horse-power at 1050 revolutions. The total weight of the machine was nearly 1200 pounds, and its speed twenty-eight miles per hour.

The eight-cylinder Curtiss motor on the *June Bug* was air cooled. This aeroplane weighed 650 pounds and made thirty-nine miles per hour, the engine developing twenty-five horse-power at 1200 turns.

RESISTANCE OF AEROPLANES

The chart on page 24 (see also the diagram of page 23) shows that the lifting power of an aeroplane increases as the angle of inclination increases, up to a certain limit. The resistance to propulsion also increases, however: and the ratio of lifting power to resistance is greatest at a very small angle — about five or six degrees. Since the motor power and weight are ruling factors in design, it is important to fly at about this angle. The supporting force is then about two pounds, and the resistance about three-tenths of a pound, per square foot of sail area, if the velocity is that assumed in plotting the chart: namely, about fifty-five miles per hour.

But the resistance R indicated on pages 23 and 24 is not the only resistance to propulsion. In addition, we have the frictional resistance of the air sliding along the sail surface. The amount of this resistance is independent of the angle of inclination: it depends directly upon the area of the planes, and in an indirect way on their dimensions in the direction of movement. It also varies nearly with the square of the velocity. At any velocity, then, the addition of this frictional resistance, which does not depend on the angle of inclination, modifies our views as to the desirable angle: and the total resistance reaches a minimum (in proportion to the weight supported) when the angle is about three degrees and the velocity about fifty miles per hour.

This is not quite the best condition, however. The skin friction does not vary exactly with the square of the velocity: and when the true law of variation is taken into account, it is found that the *horse-power* is a minimum at an angle of about five degrees and a speed of about forty miles per hour. The weight supported per horse-power may then be theoretically nearly a hundred pounds: and the frictional resistance is about one-third the direct pressure resistance. This must be regarded as the approximate condition of best effectiveness: not the exact condition, because in arriving at this result we have regarded the sails as square flat planes whereas in reality they are arched and of rectangular form.

The Question of Power

At the most effective condition, the resistance to propulsion is only about one-tenth the weight supported. Evidently the air is helping the motor.

RESISTANCE OF DIRIGIBLES

If the bow of a balloon were cut off square, its head end resistance would be that given by the rule already cited (page 19): one three-hundredth pound per square foot,

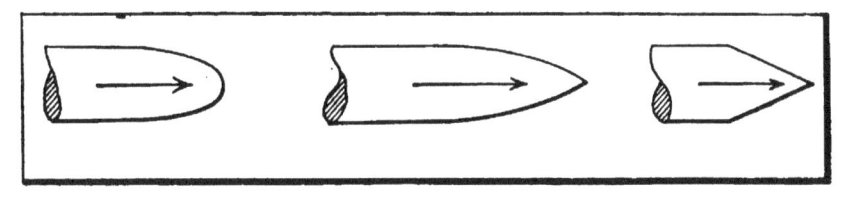

HEAD END SHAPES

multiplied by the square of the velocity. But by pointing the bow an enormous reduction of this pressure is possible. If the head end is a hemisphere (as in the English military dirigible), the reduction is about one-third. If it is a sharp cone, the reduction may be as much as four-fifths. Unless the stern is also tapered, however, there will be a considerable eddy resistance at that point.

If head end resistance were the only consideration, then for a balloon of given diameter and end shape it would be independent of the length and capacity. The longer the balloon, the better. Again, since the volume of any solid body increases more rapidly than its surface (as the linear dimensions are increased), large balloons would have a distinct advantage over small ones. The smallest dirigible ever built was that of Santos-Dumont, of about 5000 cubic feet.

Large balloons, however, are structurally weak: and more is lost by the extra bracing necessary than is gained by reduction of head end resistance. It is probable that the Zeppelin represents the limit of progress in this direction; and even in that balloon, if it had not been that the adoption of a rigid type necessitated great structural strength, it is doubtful if as great a length would have been fixed upon, in proportion to the diameter.

The frictional resistance of the air gliding along the surface of the envelope, moreover, invalidates any too arbitrary conclusions. This, as in the aeroplane, varies nearly as the square of the velocity, and is usually considerably greater than the direct head end resistance. Should the steering gear break, however, and the wind strike the *side* of the balloon, the pressure of the wind against this greatly increased area would absolutely deprive it of dirigibility.

A stationary, drifting, or "sailing" balloon may as well have the spherical as well as any other shape: it makes the wind a friend instead of a foe and requires nothing in the way of control other than regulation of altitude.

Independent Speed and Time Table

The air pressure, direct and frictional resistances, and power depend upon the *relative* velocity of flying machine and air. It is this relative velocity, not the velocity of the balloon as compared with a point on the earth's sur-

The Santos-Dumont No. 2 (1909)

face, that marks the limit of progression. Hence the speed of the wind is an overwhelming factor to be reckoned with in developing an aerial time table. If we wish to travel east at an effective speed of thirty miles per hour, while the wind is blowing due west at a speed of ten miles, our machine must have an independent speed of forty miles. On the other hand, if we wish to travel west, an independent speed of twenty miles per hour will answer.

Again, if the wind is blowing north at thirty miles per hour, and the minimum (relative) velocity at which an aeroplane will sustain its load is forty miles per hour, we cannot progress northward any more slowly than at seventy miles' speed. And we have this peculiar condition of things: suppose the wind to be blowing north at fifty miles per hour. The aeroplane designed for a forty mile speed may then face this wind and sustain itself while actually moving backward at an absolute speed (as seen from the earth) of ten miles per hour.

We are at the mercy of the wind, and wind velocities may reach a hundred miles an hour. The inherent disadvantage of aerial flight is in what engineers call its "low load factor." That is, the ratio of normal performance required to possible abnormal performance necessary under adverse conditions is extremely low. To make a balloon truly dirigible throughout the year involves, at Paris, for example, as we have seen, a speed exceeding fifty-four miles per hour: and even then, during one-tenth the year, the *effective* speed would not exceed twenty miles

IN THE BAY OF MONACO SANTOS-DUMONT'S No. 6
The flights terminated with a fall into the sea, happily without injury to the operator

per hour. A time table which required a schedule speed reduction of 60% on one day out of ten would be obviously unsatisfactory.

Further, if we aim at excessively high independent speeds for our dirigible balloons, in order to become independent of wind conditions, we soon reach velocities at which the gas bag is unnecessary: that is, a simple wing surface would at those speeds give ample support. The increased difficulty of maintaining rigidity of the envelope, and of steering, at the great pressures which would accompany these high velocities would also operate against the dirigible type.

With the aeroplane, higher speed means less sail area for a given weight and a stronger machine. Much higher speeds are probable. We have already a safe margin as to weight per horse-power of motor, and many aeroplane motors are for stanchness purposely made heavier than they absolutely need to be.

The Cost of Speed

Since the whole resistance, in either type of flying machine, is approximately proportional to the square of the velocity; and since horse-power (work) is the product of resistance and velocity, the horse-power of an air craft of any sort varies about as the cube of the speed. To increase present speeds of dirigible balloons from thirty to sixty miles per hour would then mean eight times as much horse-power, eight times as much motor weight,

eight times as rapid a rate of fuel consumption, and (since the speed has been doubled) four times as rapid a consumption of fuel in proportion to the distance traveled. Either the radius of action must be decreased, or the weight of fuel carried must be greatly increased, if higher velocities are to be attained. Present (independent) aeroplane speeds are usually about fifty miles per hour, and there is not the necessity for a great increase which exists with the lighter-than-air machines. We have already succeeded in carrying and propelling fifty pounds of total load or fifteen pounds of passenger load per horse-power of motor, with aeroplanes; the ratio of net load to horse power in the dirigible is considerably lower; but the question of weight in relation to power is of relatively smaller importance in the latter machine, where support is afforded by the gas and not by the engine.

THE PROPELLER

Very little effort has been made to utilize paddle wheels for aerial propulsion; the screw is almost universally employed. Every one knows that when a bolt turns in a stationary nut, it moves forward a distance equal to the *pitch* (lengthwise distance between two adjacent threads) at every revolution. A screw propeller is a bolt partly cut away for lightness, and the "nut" in which it works is water or air. It does not move forward quite as much as its pitch, at each revolution, because any fluid is more or less slippery as compared with a nut of solid metal. The difference between the pitch and the actual forward movement of the vessel at each revolution is called the "slip," or "slip ratio." It is never less than ten or twelve per cent in marine work, and with aerial screws is much greater. Within certain limits, the less the slip, the greater the efficiency of the propeller. Small screws have relatively greater slips and less efficiency, but are lighter. The maximum efficiency of a screw propeller in water is under 80%. According to Langley's experiments, the usual efficiency in air is only about 50%. This means that only half the power of the motor will be actually available for producing forward movement — a conclusion already foreshadowed.

In common practice, the pitch of aerial screws is not far from equal to the diameter. The rate of forward movement, if there were no slip, would be proportional to the pitch and the number of revolutions per minute. If the latter be increased, the former may be decreased. Screws direct-connected to the motors and running at high speeds will therefore be of smaller pitch and diameter than those run at reduced speed by gearing, as in the machine illustrated on page 134. The number of blades is usually two, although this gives less perfect balance than would a larger number. The propeller is in many monoplanes placed in *front*: this interferes, unfortunately, with the air currents against the supporting surfaces.

There is always some loss of power in the bearings and power-transmitting devices between the motor and propeller. This may decrease the power usefully exerted even to *less* than half that developed by the motor.

GETTING UP AND DOWN: MODELS AND GLIDERS: AEROPLANE DETAILS

LAUNCHING

THE Wright machines (at least in their original form) have usually been started by the impetus of a falling weight, which propels them along skids until the velocity suffices to produce ascent. The preferred designs among French machines have contemplated self-starting equipment.

WRIGHT BIPLANE ON STARTING RAIL, SHOWING PYLON AND WEIGHT

This involves mounting the machine on pneumatic-tired bicycle wheels so that it can run along the ground. If a fairly long stretch of good, wide, straight road is available, it is usually possible to ascend. The effect of altitude and atmospheric density on sustaining power is forcibly illustrated by the fact that at Salt Lake City one of the aviators was unable to rise from the ground.

To accelerate a machine from rest to a given velocity in a given time or distance involves the use of propulsive

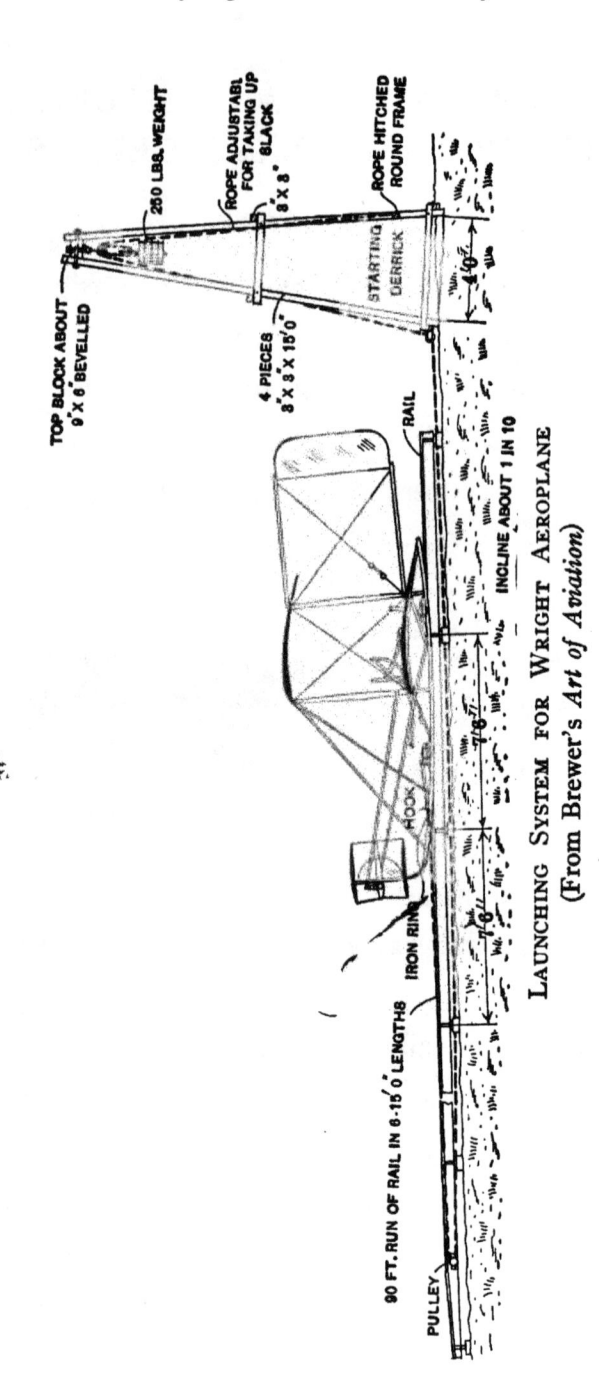

LAUNCHING SYSTEM FOR WRIGHT AEROPLANE
(From Brewer's *Art of Aviation*)

force additional to that necessary to maintain the velocity attained. Apparently, therefore, any self-starting machine must have not only the extra weight of framework and wheels but also extra motor power.

Upon closer examination of the matter, we may find a particularly fortunate condition of things in the aeroplane. Both sustaining power and resistance vary with the inclination of the planes, as indicated by the chart on page 24. It is entirely possible to start with no such inclination, so that the direct wind resistance is eliminated. The motor must then overcome only air friction, in addition to providing an accelerating force. The machine runs along the ground, its velocity rapidly increasing. As soon as the necessary speed (or one somewhat greater) is attained, the planes are tilted and the aeroplane rises from the ground.

The velocity necessary to just sustain the load at a given angle of inclination is called the *critical* or *soaring* velocity. For a given machine, there is an angle of inclination (about half a right angle) at which the minimum speed is necessary. This speed is called the "least soaring velocity." If the velocity is now increased, the angle of inclination may be reduced and the planes will soar through the air almost edgewise, apparently with diminished resistance and power consumption. This decrease in power as the speed increases is called *Langley's Paradox*, from its discoverer, who, however, pointed out that the rule does not hold in practice when frictional resistances are

THE NIEUPORT MONOPLANE
Self-Starting with an 18 hp. motor (From *The Air Scout*)

included. We cannot expect to actually save power by moving more rapidly than at present; but we should have to provide much more power if we tried to move much more slowly.

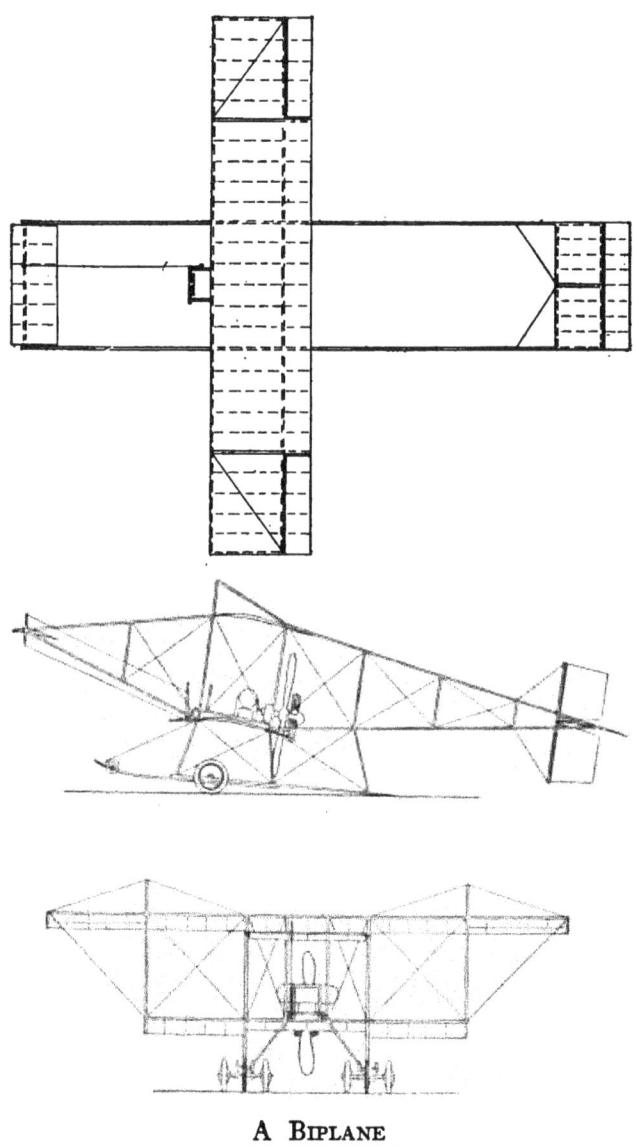

A Biplane
(From *Aircraft*)

Economical and practicable starting of an aeroplane thus requires a free launching space, along which the machine may accelerate with nearly flat planes: a downward slope would be an aid. When the planes are tilted for ascent, after attaining full speed, quick control is necessary to avoid the possibility of a back-somersault. A fairly wide

(Photo by American Press Association)
ELY AT LOS ANGELES

launching platform of 200 feet length would ordinarily suffice. The flight made by Ely in January of this year, from San Francisco to the deck of the cruiser *Pennsylvania* and back, demonstrated the possibility of starting from a limited area. The wooden platform built over the after deck of the warship was 130 feet long, and sloped. On the return trip, the aeroplane ran down this slope, dropped somewhat, and then ascended successfully.

If the effort is made to ascend at low velocities, then the motor power must be sufficient to propel the machine at an extreme angle of inclination — perhaps the third of a right angle, approximating to the angle of least velocity for a given load. According to Chatley, this method of starting by Farman at Issy-les-Molineaux involved the use of a motor of fifty horse-power: while Roe's machine at Brooklands rose, it is said, with only a six horse-power motor.

DESCENDING

What happens when the motor stops? The velocity of the machine gradually decreases: the resistance to forward movement stops its forward movement and the

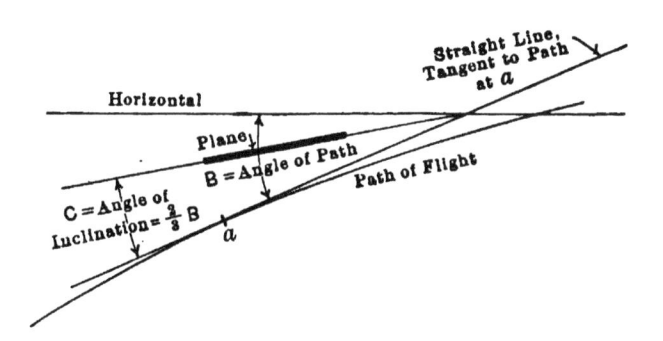

excess of weight over upward pressure due to velocity causes it to descend. It behaves like a projectile, but the details of behavior are seriously complicated by the variation in head resistance and sustaining force due to

Descending

changes in the angle of the planes. The "angle of inclination" is now not the angle made by the planes with the horizontal, but the angle which they make with the path of flight. Theory indicates that this should be about two-thirds the angle which the path itself makes with the horizontal: that is, the planes themselves are inclined downward toward the front. The forces which determine the descent are fixed by the velocity and the angle between the planes and the path of flight. Manipulation of the rudders and main planes or even the motor may be practised to ensure lancing to best advantage; but in spite of these (or perhaps on account of these) scarcely any part of aviation offers more dangers, demands more genius on the part of the operator, and has been less satisfactorily analyzed than the question of "getting down." It is easy to stay up and not very hard to "get up," weather conditions being favorable; but it is an "all-sufficient job" to *come down*. Under the new rules of the International Aeronautic Federation, a test flight for a pilot's license must terminate with a descent (motor stopped) in which the aviator is to land within fifty yards of the observers and come to a full stop inside of fifty yards therefrom. The elevation at the beginning of descent must be at least 150 feet.

GLIDERS

If the motor and its appurtenances, and some of the purely auxiliary planes, be omitted, we have a *glider*. The glider is not a toy; some of the most important problems of balancing may perhaps be some day solved by its aid. Any boy may build one and fly therewith, although a large kite promises greater interest. The cost is trifling, if the framework is of bamboo and the surfaces are cotton. Areas of glider surfaces frequently exceed 100 square feet. This amount of surface is about right for a person of mod-

THE WITTEMAN GLIDER

erate weight if the machine itself does not weigh over fifty pounds. By running down a slope, sufficient velocity may be attained to cause ascent; or in a favorable wind (up the slope) a considerable backward flight may be experienced. Excessive heights have led to fatal accidents in gliding experiments.

Models

The building of flying models has become of commercial importance. It is not difficult to attain a high ratio of surface to weight, but it is almost impossible to get motor power in the small units necessary without exceeding the permissible limit of motor weight. No gasoline engine or electric motor can be made sufficiently light for a toy model. Clockwork springs, if especially designed, may give the necessary power for short flights, but no better form of power is known just now than the twisted rubber band. For the small boy, a biplane with sails about eighteen inches by four feet, eighteen inches apart, anchored under his shoulders by six-foot cords while he rides his bicycle, will give no small amount of experience in balancing and will support enough of a load to make the experiment interesting.

Some Details: Balancing

It is easily possible to compute the areas, angles, and positions of auxiliary planes to give desired controlling or stabilizing effects; but the computation involves the use of accurate data as to positions of the various weights, and on the whole it is simpler to correct preliminary calculations by actually supporting the machine at suitable points and observing its balance. Stability is especially uncertain at very small angles of inclination, and such angles are to be avoided whether in ordinary operation or in descent. The necessity for rotating main planes in order to produce ascent is disadvantageous on this ground; but the proposed use of sliding or jockey weights for supple-

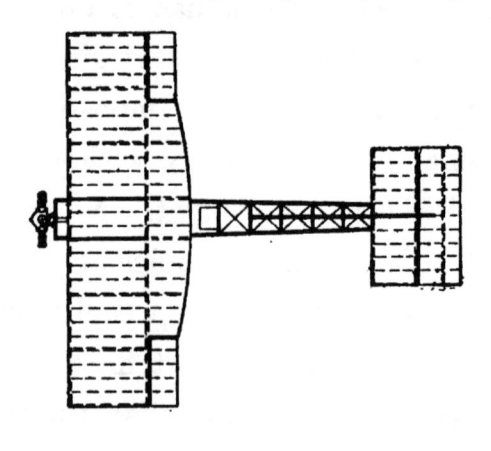

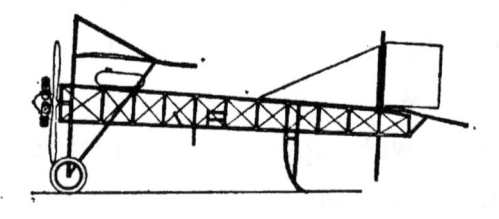

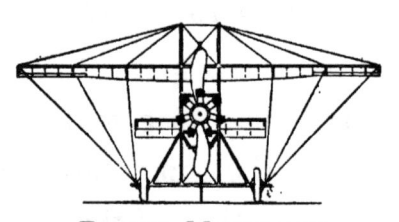

FRENCH MONOPLANE
(From *Aircraft*)

mentary balancing appears to be open to objections no less serious. Steering may be perceptibly assisted, in as delicately a balanced device as the aeroplane, by the inclination of the body of the operator, just as in a bicycle. The direction of the wind in relation to the required course may seriously influence the steering power. Suppose the course to be northeast, the wind east, the independent speed of the machine and that of the wind being the same. The car will head due north. By bringing the rudder in position (*a*), the course may be changed to north, or nearly so, the wind exerting a powerful pressure on the

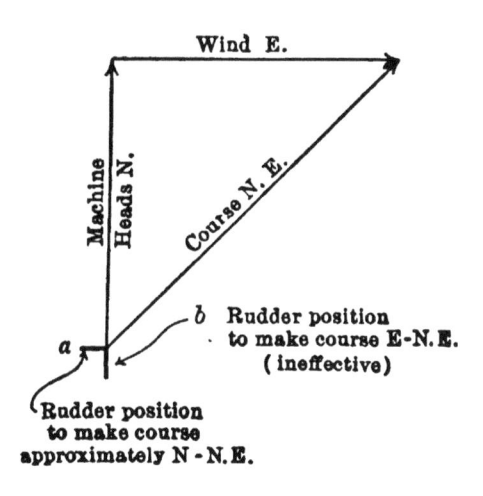

Wind E.

Machine Heads N.

Course N.E.

b Rudder position to make course E-N.E. (ineffective)

a

Rudder position to make course approximately N-N.E.

rudder; but if a more easterly or east-northeast course be desired, and the rudder be thrown into the usual position therefor (*b*), it will exert no influence whatever, because it is moving before the wind and precisely at the speed of the wind.

It might be thought that, following analogies of marine engineering, the center of gravity of an aeroplane should be kept low. The effect of any unbalanced pressure or force against the widely extended sails of the machine is to rotate the whole apparatus about its center of gravity.

Lejeune Biplane (385 lbs., 10-12 hp.)

The further the force from the center of gravity, the more powerful is the force in producing rotation. The defect in most aeroplanes (especially biplanes) is that the center of gravity is *too* low. If it could be made to coincide with the center of disturbing pressure, there would be no unbalancing effect from the latter. It is claimed that the steadiest machines are those having a high center of gravity; and the claim, from these considerations, appears reasonable.

WEIGHTS

It has been found not difficult to keep down the weight of framework and supporting surfaces to about a pound per square foot. The most common ratio of surface to

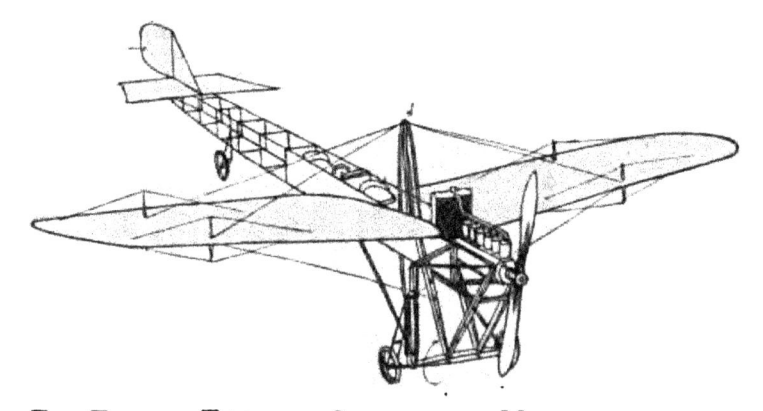

THE TELLIER TWO-SEAT SIX-CYLINDER MONOPLANE AT THE PARIS SHOW
One of this type has been sold to the Russian Government
(From *Aircraft*)

total weight is about one to two: so that the machinery and operator will require one square foot of surface for each pound of their weight. On this basis, the smallest possible man-carrying aeroplane would have a surface scarcely below 250 square feet. Most biplanes have twice this surface: a thousand square feet seems to be the limit without structural weakness. Some recent French machines, designed for high speeds, show a greatly increased ratio of weight to surface. The *Hanriot*, a monoplane with wings upwardly inclined toward the outer edge, carries over 800 pounds on less than 300 square feet. The Farman monoplane of only 180 square feet sustains over 600 pounds. The same aviator's racing biplane is stated to support nearly 900 pounds on less than 400 square feet.

Motor weights can be brought down to about two pounds per horse-power, but such extreme lightness is not always needed and may lead to unreliability of operation. The effect of an accumulation of ice, sleet, snow, rain, or dew might be serious in connection with flights in high altitudes or during bad weather. After one of his last year's flights at Étampes Mr. Farman is said to have descended with an extra load of nearly 200 pounds on this account. With ample motor power, great flexibility in weight sustention is made possible by varying the inclination of the planes. In January of this year, Sommer at Douzy carried six passengers in a large biplane on a cross-country flight: and within the week afterward a monoplane operated by Le Martin flew for five minutes with the aeronaut and seven passengers, at Pau. The total weight lifted was about half a ton, and some of the passengers must have been rather light. The two-passenger Fort Myer biplane of the Wright brothers is understood to have carried about this total weight. These records have, how-

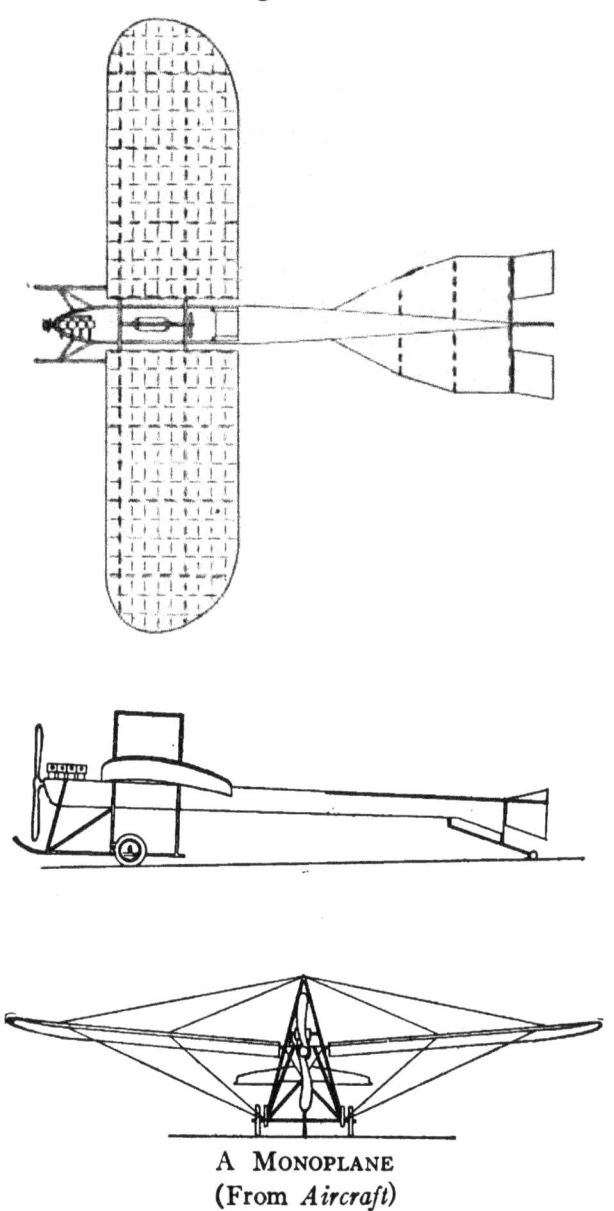

A Monoplane
(From *Aircraft*)

ever, been surpassed since they were noted. Breguet, at Douai, in a deeply-arched biplane of new design, carried eleven passengers, the total load being 2602 pounds, and that of aeronaut and passengers alone 1390 pounds. The flight was a short one, at low altitude; but the same aviator last year made a long flight with five passengers, and carried a load of 1262 pounds at 62 miles per hour. And as if in reply to this feat, Sommer carried a live load of 1436 pounds (13 passengers) for nearly a mile, a day or two later, at Mouzon One feels less certain than formerly, now, in the snap judgment that the heavier-than-air machine will never develop the capacity for heavy loads.

Miscellaneous

French aviators are fond of employing a carefully designed car for the operator and control mechanism. The Wright designs practically ignore the car: the aviator sits on the forward edge of the lower plane with his legs hanging over.

It has been found that auxiliary planes must not be too close to the main wings: a gap of a distance about 50% greater than the width of the widest adjacent plane must be maintained if interference with the supporting air currents is to be avoided. Main planes are now always arched; auxiliary planes, not as universally. The concave under surface of supporting wings has its analogy in the wing of the bird and had long years since been applied in the parachute.

The car (if used) and all parts of the framework should be of "wind splitter" construction, if useless resistance is

Getting Up and Down

to be avoided. The ribs and braces of the frame are of course stronger, weight for weight, in this shape, since a

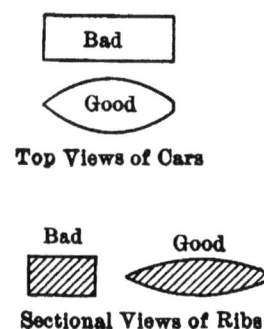

Bad

Good

Top Views of Cars

Bad   Good

Sectional Views of Ribs

narrow deep beam is always relatively stronger than one of square or round section. Excessive frictional resistance is to be avoided by using a smoothly finished fabric for the wings, and the method of attaching this fabric

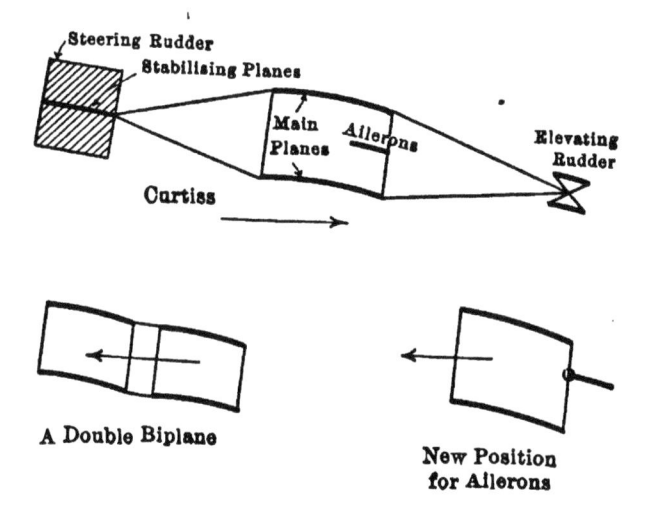

Steering Rudder
Stabilising Planes
Main Planes Ailerons   Elevating Rudder
Curtiss

A Double Biplane

New Position for Ailerons

to the frame should be one that keeps it as flat as possible at all joints.

The sketches give the novel details of some machines recently exhibited at the Grand Central Palace in New

York. The stabilizing planes were invariably found in the rear, in all machines exhibited.

THE THINGS TO LOOK AFTER

The operator of an aeroplane has to do the work of at least two men. No vessel in water would be allowed to attain such speeds as are common with air craft, unless provided with both pilot and engineer. The aviator is his own pilot and his own engineer. He must both manage his propelling machinery and steer. Separate control for vertical rudders, elevating rudders and ailerons, for starting the engine; the adjustment of the carbureter, the spark, and the throttle to get the best results from the motor; attention to lubrication and constant watchfulness of the water-circulating system: these are a few of the things for him to consider; to say nothing of the laying of his course and the necessary anticipation of wind and altitude conditions.

These things demand great resourcefulness, but — for their best control — involve also no small amount of scientific knowledge. For example, certain adjustments at the motor may considerably increase its power, a possibly necessary increase under critical conditions: but if such adjustments also decrease the motor efficiency there must be a nice analysis of the two effects so that extra power may not be gained at too great a cost in radius of action.

The whole matter of flight involves both sportsman's and engineer's problems. Wind gusts produce the same

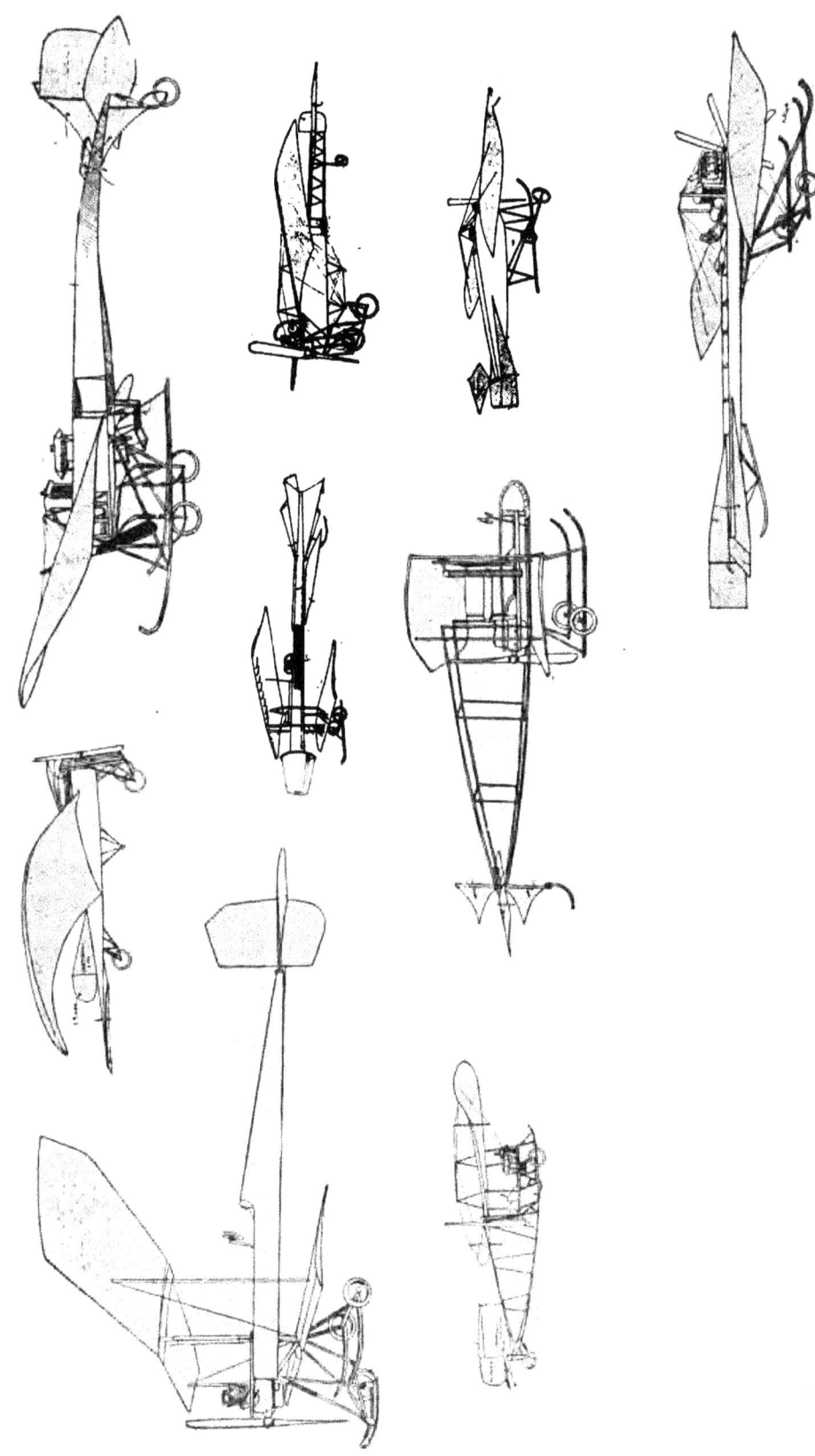

Some Recent French Machines (From *Aircraft*)

effects as "turning corners"; or worse — rapidly changing the whole balance of the machines and requiring immediate action at two or three points of control. Both ascent and descent are influenced by complicated laws and are scarcely rendered safe — under present conditions — by the most ample experience. A lateral air current bewilders the steering and also demands special promptness and skill. To avoid disturbing surface winds, even over open country, a minimum flying height of 300 feet is considered necessary. This height, furthermore, gives more choice in the matter of landing ground than a lower elevation.

When complete and automatic balance shall have been attained — as it must be attained — we may expect to see small amateur aeroplanes flying along country roads at low elevations — perhaps with a guiding wheel actually in contact with the ground. They will cost far less than even a small automobile, and the expense for upkeep will be infinitely less. The grasshopper will have become a water-spider.

SOME AEROPLANES — SOME ACCOMPLISHMENTS

THE Wright biplane has already been shown (see pages 31, 37, 121, 122). It was distinguished by the absence

ORVILLE WRIGHT AT FORT MYER, VA., 1908

of a wheel frame or car and by the wing-warping method of stabilizing. Later Wright machines have the spring frame and wheels for self-starting. The best known aeroplane of this design was built to meet specifications of the United

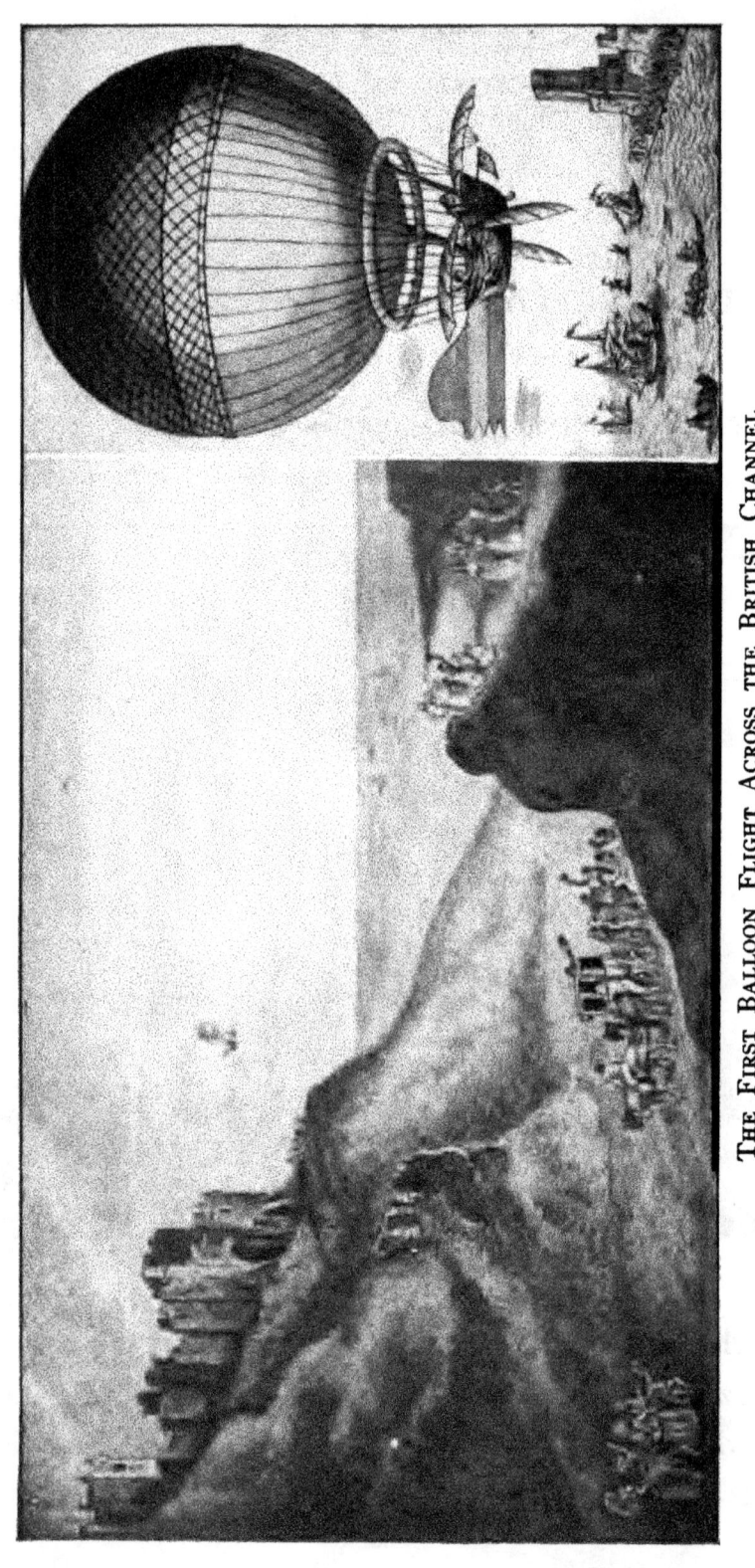

THE FIRST BALLOON FLIGHT ACROSS THE BRITISH CHANNEL
More than a century before Blériot's feat, Blanchard crossed from Dover to Calais

Some Aeroplanes — Some Accomplishments 145

States Signal Corps issued in 1907. It was tried out during 1908 at Fort Myer, Va., while one of the Wright brothers was breaking all records in Europe: making

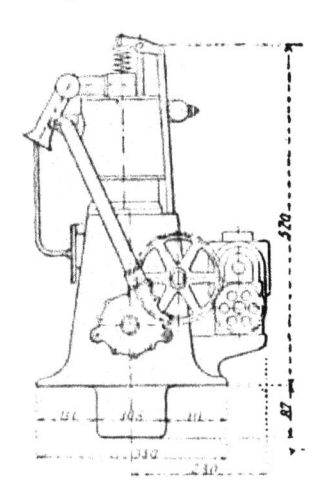

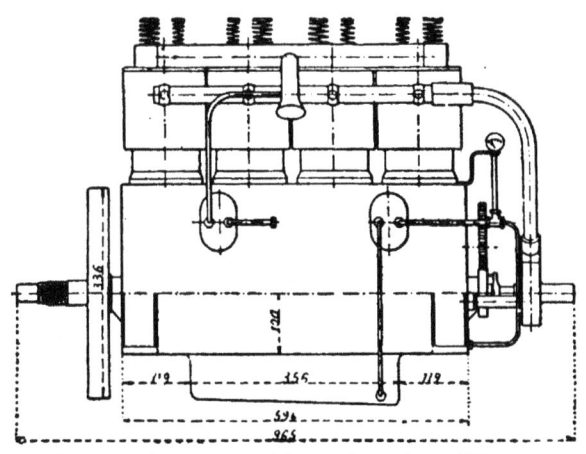

WRIGHT MOTOR. Dimensions in millimeters
(From Petit's *How to Build an Aeroplane*)

over a hundred flights in all, first carrying a passenger and attaining the then highest altitude (360 feet) and greatest distance of flight (seventy-seven miles).

The ownership of the Wrights in the wing-warping

method of control is still the subject of litigation. The French infringers, it is stated, concede priority of application to the Wright firm, but maintain that such publicity was given the device that it was in general use before it was patented.

The Fort Myer machine had sails of forty feet spread, six and one-half feet deep, with front elevating planes three by sixteen feet. It made about forty miles per hour with two passengers. The apparatus was specified to carry a passenger weight of 350 pounds, with fuel for a 125-mile flight. The main planes were six feet apart. The steering rudder (double) was of planes one foot deep and nearly six feet high. The four-cylinder-four-cycle, water-cooled motor developed twenty-five horse-power at 1400 revolutions. The two propellers, eight and one-half feet in diameter, made 400 revolutions.

The flight by Mr. Wilbur Wright from the Statue of Liberty to the tomb of General Grant, in New York, 1909, and the exploits of his brother in the same year, when a new altitude record of 1600 feet was made and H.R.H. the Crown Prince of Germany was taken up as a passenger, are only specimens of the later work done by these pioneers in aerial navigation.

Like the Wrights, the Voisin firm from the beginning adhered firmly to the biplane type of machine. The sketch gives dimensions of one of the early cellular forms built for H. Farman (see illustration, page 147). The metal screw makes about a thousand revolutions. The

Some Aeroplanes — Some Accomplishments 147

wings are of india rubber sheeting on an ash frame, the whole frame and car body being of wood, the latter covered with canvas and thirty inches wide by ten feet long. The engine weighed 175 pounds. The whole weight of this machine was nearly 1200 pounds; that built later for Delagrange was brought under a thousand pounds. The ratio of weight to main surface in the Farman aeroplane was about $2\frac{3}{4}$ to 1.

A modified cellular biplane also built for Farman had a main wing area of 560 square feet, the planes being sev-

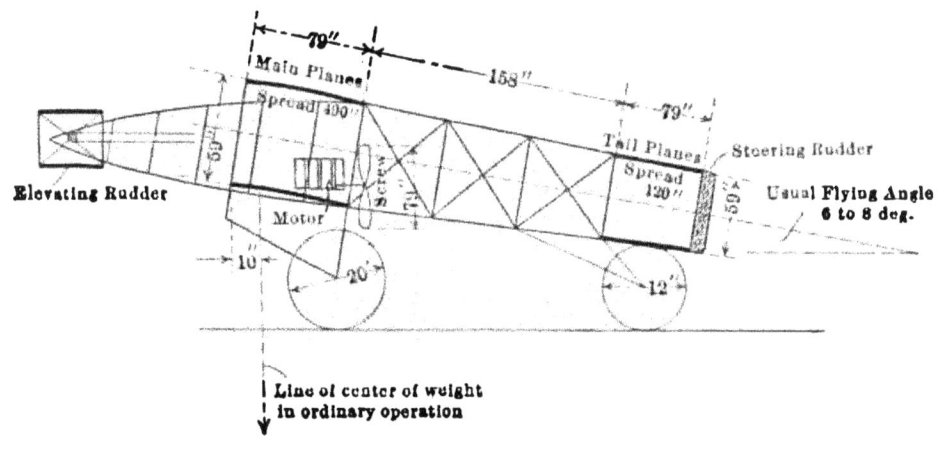

VOISIN-FARMAN BIPLANE

enty-nine inches wide and only fifty-nine inches apart. The tail was an open box, seventy-nine inches wide and of about ten feet spread. The cellular partitions in this tail were pivoted along the vertical front edges so as to serve as steering rudders. The elevating rudder was in front. The total weight was about the same as that of the first machine and the usual speed twenty-eight miles per hour.

Henry Farman has been flying publicly since 1907. He

made the first circular flight of one kilometer, and attained a speed of about a mile a minute, in the year following.

THE CHAMPAGNE GRAND PRIZE WON BY HENRY FARMAN
80 Kilometers in 3 hours

In 1909 he accomplished a trip of nearly 150 miles, remaining four hours in the air. Farman was probably the first man to ascend with two passengers.

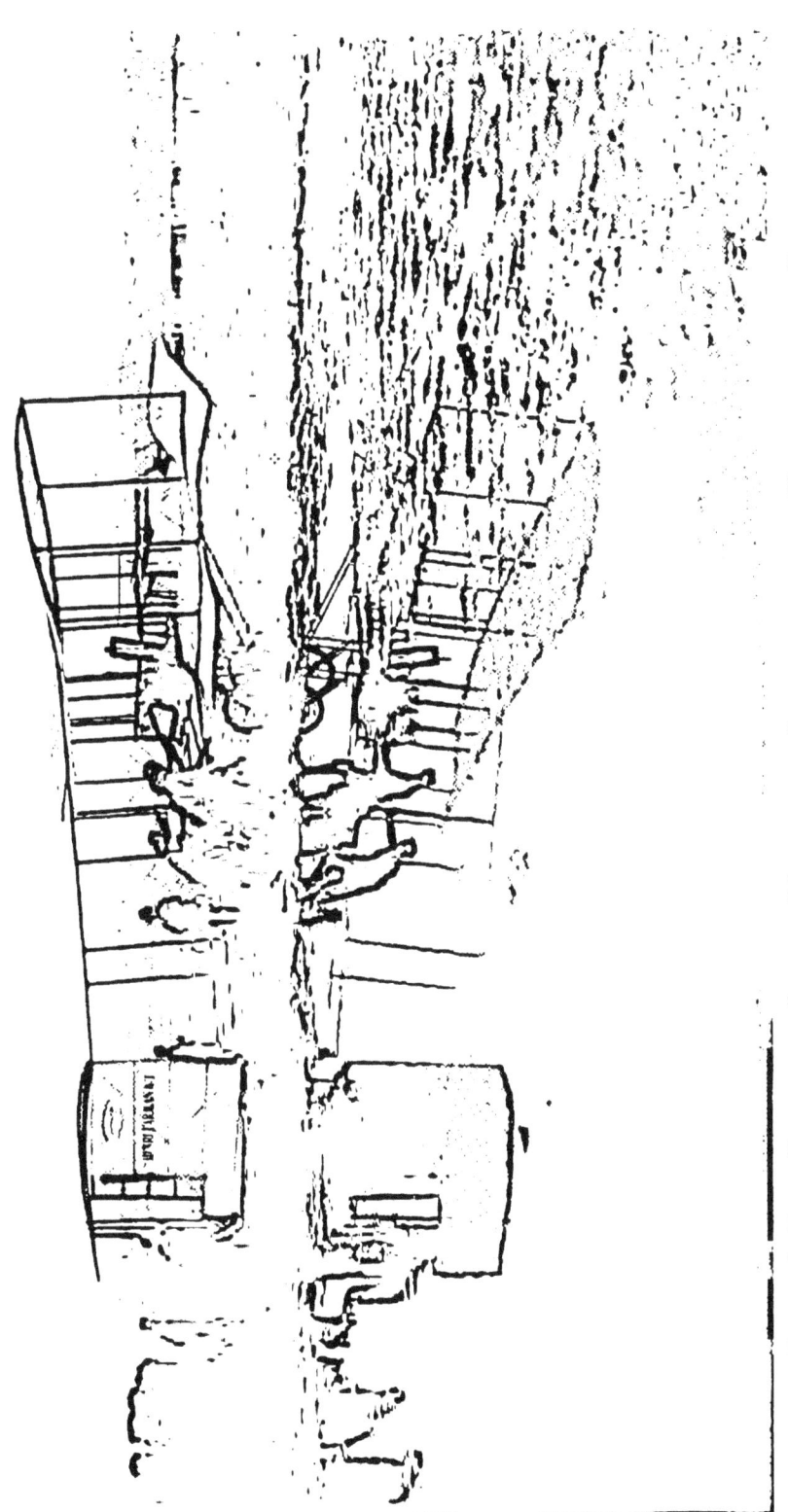

Farman's First Biplane at Issy-les-Moulineaux Returning to the Hangar after a Flight

The *June Bug*, one of the first Curtiss machines, is shown below. This was one of the lightest of biplanes, having a wing spread of forty-two feet and an area of 370 square feet. The wings were transversely arched, being furthest apart at the center: an arrangement which has not been continued. It had a box tail, with a steering rudder of about six square feet area, *above* the tail. The horizontal rudder, in front, had a surface of twenty

THE "JUNE BUG"

square feet. Four triangular ailerons were used for stability. The machine had a landing frame and wheels, made about forty miles per hour, and weighed, in operation, 650 pounds.

Mr. Curtiss first attained prominence in aviation circles by winning the *Scientific American* cup by his flight at the speed of fifty-seven miles per hour, in 1908. In the following year he exhibited intricate curved flights at Mineola, and circled Governor's Island in New York harbor. In 1910 he made his famous flight from Albany to New York, stopping *en route*, as prearranged. At Atlantic City he flew fifty miles over salt water. A flight of seventy miles over Lake Erie was accomplished in September of the same year, the return trip being made the following day. On January 26, 1911, Curtiss repeatedly

(Photo by Levick, N.Y.)
CURTISS BIPLANE

ascended and descended, with the aid of hydroplanes, in San Diego bay, California: perhaps one of the most important of recent achievements. It is understood that Mr. Curtiss is now attempting to duplicate some of these performances under the high-altitude conditions of Great Salt Lake. According to press reports, he has been invited to give a similar demonstration before the German naval authorities at Kiel.

The *aeroscaphe* of Ravard was a machine designed to move either on water or in air. It was an aeroplane with

CURTISS' HYDRO-AEROPLANE AT SAN DIEGO GETTING UNDER WAY
(From the *Columbian Magazine*)

pontoons or floaters. The supporting surface aggregated 400 square feet, and the gross weight was about 1100 pounds. A fifty horse-power Gnome seven-cylinder motor at 1200 revolutions drove two propellers of eight and ten and one-half feet diameter respectively: the propellers being mounted one behind the other on the same shaft.

Ely's great shore-to-warship flight was made without the aid of the pontoons which he carried. Ropes were stretched across the landing platform, running over sheaves and made fast to heavy sand bags. As a further precau-

FLYING OVER THE WATER AT FIFTY MILES PER HOUR
Curtiss at San Diego Bay
(From the *Columbian Magazine*)

tion, a canvas barrier was stretched across the forward end of the platform. The descent brought the machine to the platform at a distance of forty feet from the upper end: grappling hooks hanging from the framework of the aeroplane then caught the weighted ropes, and the speed was checked (within about sixty feet) so gradually that "not a wire or bolt of the biplane was injured."

BLÉRIOT-VOISIN CELLULAR BIPLANE WITH PONTOONS
Hauled by a Motor Boat

Latham's "Antoinette"

156 *Flying Machines Today*

JAMES J. WARD AT LEWISTON FAIR, IDAHO
Flying Machine Mfg. Co. Biplane (30 hp. Motor)

Some Aeroplanes — Some Accomplishments 157

MARCEL PENOT IN THE MOHAWK BIPLANE, Mineola to Hicksville, L. I. 26 miles cross-country in 30 minutes (50 hp. Harriman Engine)

Recent combinations of aeroplane and automobile, and aeroplane with motor boat, have been exhibited. One of the latter devices is like any monoplane, except that the lower part is a water-tight aluminum boat body carrying three passengers. It is expected to start of itself from the water and to fly at a low height like a flying fish at a speed of about seventy-five miles per hour. Should anything go wrong, it is capable of floating on the water.

In the San Diego Curtiss flights, the machine skimmed along the surface of the bay, then rose to a height of a hundred feet, moved about two miles through the air in a circular course, and finally alighted close to its starting-point in the water. Turns were made in water as well as in air, a speed of forty miles per hour being attained while "skimming." The "hydroplanes" used are rigid flat surfaces which utilize the pressure of the water for sustention, just as the main wings utilize air pressure. On account of the great density of water, no great amount of surface is required: but it must be so distributed as to balance the machine. The use of pontoons makes it possible to rest upon the water and to start from rest. A trip like Ely's could be made without a landing platform, with this type of machine; the aeroplane could either remain alongside the war vessel or be hoisted aboard until ready to venture away again.

There are various other biplanes attracting public attention in this country. In France the tendency is all toward

Some Aeroplanes — Some Accomplishments 159

the monoplane form, and many of the "records" have, during the past couple of years, passed from the former to

SANTOS-DUMONT'S "DEMOISELLE"

the latter type of machine. The monoplane is simpler and usually cheaper. The biplane may be designed for greater economy in weight and power. Farman has lately experimented with the monoplane type of machine: the large number of French designs in this class discourages any attempt at complete description.

The smallest of aeroplanes is the Santos-Dumont *Demoiselle*. The original machine is said to have supported 260 pounds on 100 square feet of area, making a speed of sixty miles per hour. Its proprietor was the first aviator

BLÉRIOT MONOPLANE

in Europe of the heavier-than-air class. After having done pioneer work with dirigible balloons, he won the Deutsch prize for a hundred meter aeroplane flight (the first outside of the United States) in 1906; the speed being twenty-three miles per hour. His first flight, of 400 feet, in a monoplane was made in 1907.

The master of the monoplane has been Louis Blériot. Starting in 1907 with short flights in a Langley type of

Some Aeroplanes — Some Accomplishments 161

machine, he made his celebrated cross-country run, and the first circling flights ever achieved in a monoplane, the

LATHAM'S FALL INTO THE CHANNEL

following year. On July 25, 1909, he crossed the British Channel, thirty-two miles, in thirty-seven minutes.

The Channel crossing has become a favorite feat. Mr.

Latham, only two days after Blériot, all but completed it in his Antoinette monoplane. De Lesseps, in a Blériot machine, was more fortunate. Sopwith, last year, won the de Forest prize of $20,000 by a flight of 174 miles from England into Belgium. The ill-fated Rolls made the round trip between England and France. Grace, contesting for the same prize, reached Belgium, was driven back to Calais, started on the return voyage, and vanished — all save some few doubtful relics lately found. Moisant reached London from Paris — the first trip on record between these cities without change of conveyance: and one which has just been duplicated by Pierre Prier, who, on April 12, made the London to Paris journey, 290 miles, in 236 minutes, without a stop. This does not, however, make the record for a continuous flight: which was attained by Tabuteaw, who at Buc, on Dec. 30, 1910, flew around the aerodrome for 465 minutes at the speed of $48\frac{1}{2}$ miles per hour.

Other famous crossings include those of the Irish Sea, 52 miles, by Loraine; Long Island Sound, 25 miles, by Harmon; and Lake Geneva, 40 miles, by Defaux.

It was just about a century ago that Cayley first described a soaring machine, heavier than air, of a form remarkably similar to that of the modern aeroplane. Aside from Henson's unsuccessful attempt to build such a machine, in 1842, and Wenham's first gliding experiments with a triplane in 1857, soaring flight made no real progress until Langley's experiments. That investigator, with

DE LESSEPS IN A BLÉRIOT CROSSING THE CHANNEL

(Photo by Levick, N.Y.)

Maxim and others, ascertained those laws of aerial sustention the application of which led to success in 1903.

The eight years since have held the crowded hours of aviation. Before this book is printed, it may be rendered obsolete by new developments. The exploits of Paulhan, of R. E. Peltèrie since 1907, Bell's work with his tetrahedral kites — all have been either stimulating or directly fruitful. Delagrange began to break speed records in 1908.

THE MAXIM AEROPLANE

A year later he attained a speed of fifty miles. The first woman to enjoy an aeroplane voyage was Mme. Delagrange, in Turin, in 1908.

The first flight in England by an English-built machine was made in January, 1909. That year, Count de Lambert flew over Paris, and in 1910 Grahame-White circled his machine over the city of Boston. The year 1910 sur-

LANGLEY'S AEROPLANE (1896)
Steam driven

passed all its predecessors in increasing the range and control of aeroplanes; over 1500 ascents were made by Wright machines alone; but 1911 promises to show even greater results. Three men made cross-country flights from Belmont Park to the Statue of Liberty and back,

ROBART MONOPLANE.

in New York;* at least five men attained altitudes exceeding 9,000 feet. Hamilton made the run from New York to Philadelphia and return, in June. The unfortunate Chavez all but abolished the fames of Hannibal and Napo-

* The contestants for the Ryan prize of $10,000 were Moisant, Count de Lesseps, and Grahame-White. Owing to bad weather, there was no general participation in the preliminary qualifying events, and some question exists as to whether such qualification was not tacitly waived; particularly in view of the fact that the prize was awarded to the technically unqualified competitor, Mr. Moisant, who made the fastest time. This award was challenged by Mr. Grahame-White, and upon review by the International Aeronautic Federation the prize was given to de Lesseps, the slowest of the contestants, Grahame-White being disqualified for having fouled a pylon at the start. This gentleman has again appealed the case, and a final decision cannot be expected before the meeting of the Federation in October, 1911.

leon by crossing the icy barrier of the Alps, from Switzerland to Italy — in forty minutes!

Tabuteau, almost on New Year's eve, broke all distance records by a flight of 363 miles in less than eight hours; while Barrier at Memphis probably reached a speed of eighty-eight miles per hour (timing unofficial). With the new year came reports of inconceivable speeds by a machine skidding along the ice of Lake Erie; the successful

VINA MONOPLANE

receipt by Willard and McCurdy of wireless messages from the earth to their aeroplanes; and the proposal by the United States Signal Corps for the use of flying machines for carrying Alaskan mails.

McCurdy all but succeeded in his attempt to fly from Key West to Havana, surpassing previous records by remaining aloft above salt water while traveling eighty miles. Lieutenant Bague, in March, started from Antibes, near Nice, for Corsica. After a 124-mile flight, breaking all records for sea journeys by air, he reached the islet of Gorgona, near Leghorn, Italy, landing on bad ground and badly damaging his machine. The time of flight was $5\frac{1}{2}$ hours. Bellinger completed the 500-mile "accommodation train" flight from Vincennes to Pau; Vedrine, on April 12, by making the same journey in 415 minutes of actual flying time, won the Béarn prize of $4000; Say attained a speed of 74 miles per hour in circular flights at Issy-les-Moulineaux. Aeroplane flights have been made in Japan, India, Peru, and China.

One of the most spectacular of recent achievements is that of Renaux, competing for the Michelin Grand Prize. A purse of $20,000 was offered in 1909 by M. Michelin, the French tire manufacturer, for the first successful flight from Paris to Clermont-Ferrand — 260 miles — in less than six hours. The prize was to stand for ten years. It was prescribed that the aviator must, at the end of the journey, circle the tower of the Cathedral and alight on the summit of the Puy de Dome — elevation 4500 feet — on a landing place measuring only 40 by 100 yards, surrounded by broken and rugged ground and usually obscured by fog.

The flight was attempted last year by Weymann, who fell short of the goal by only a few miles. Leon Morane met with a serious accident, a little later, while attempting the trip with his brother as a passenger. Renaux completed the journey with ease in his Farman biplane, carrying a passenger, his time being 308 minutes.

This Michelin Grand Prize is not to be confused with the

Michelin Trophy of $4000 offered yearly for the longest flight in a closed circuit.

Speeds have increased 50% during the past year; even with passengers, machines have moved more than a mile a minute: average motor capacities have been doubled or tripled. The French men and machines hold the records for speed, duration, distance, and (perhaps) altitude. The highest altitude claimed is probably that attained by Garros at Mexico City, early this year — 12,052 feet above sea level. The world's speed record for a two-man flight appears to be that of Foulois and Parmalee, made at Laredo, Texas, March 3, 1911: 106 miles, cross-country, in 127 minutes. Three-fourths of all flights made up to this time have been made in France — a fair proportion, however, in American machines.

NOTE

The rapidity with which history is made in aeronautics is forcibly suggested by the revision of text made necessary by recent news. The new *Deutschland* has met the fate of its predecessors; the Paris–Rome–Turin flight is at this moment under way; and Lieutenant Bayne, attempting once more his France–to–Corsica flight, has — for the time being at least — disappeared.

THE POSSIBILITIES IN AVIATION

MEN now fly and will probably keep on flying; but aviation is still too hazardous to become the popular sport of the average man. The overwhelmingly important problem with the aeroplane is that of stability. These machines must have a better lateral balance when turning corners or when subjected to wind gusts: and the balance must be automatically, not manually, produced.

BLANC MONOPLANE

Other necessary improvements are of minor urgency and in some cases will be easy to accomplish. Better mechanical construction, especially in the details of attachments, needs only persistence and common sense. Structural strength will be increased; the wide spread of wing presents difficulties here, which may be solved either by increasing the number of superimposed surfaces, as in triplanes, or in some other manner. Greater carrying capac-

MELVIN VANIMAN TRIPLANE

ity — two men instead of one — may be insisted upon; and this leads to the difficult question of motor weights. The revolving air-cooled motor may offer further possibil-

JEAN DE CRAWHEZ TRIPLANE

ities: the two-cycle idea will help if a short radius of action is permissible: but a weight of less than two pounds to

172 *Flying Machines Today*

the horse-power seems to imply, almost essentially, a lack of ruggedness and surety of operation. A promising field

A Triplane

for investigation is in the direction of increasing propeller efficiencies. If such an increase can be effected, the whole of the power difficulty will be greatly simplified.

This same motor question controls the proposal for increased speed. The use of a reserve motor would again increase weights; though not necessarily in proportion to the aggregate engine capacity. Perhaps something may be accomplished with a gasoline turbine, when one is developed. In any case, no sudden increase in speeds seems to be probable; any further lightening of motors must be undertaken with deliberation and science. If much higher maximum speeds are attained, there will be an opportunity to vary the speed to suit the requirements. Then clutches, gears, brakes, and speed-changing devices of various sorts will become necessary, and the problem of weights of journal bearings — already no small matter — will be made still more serious. And with variable speed must probably come variable sail area — in preference to tilting — so that the fabric must be reefed on its frame. Certainly two men, it would seem, will be needed!

Better methods for starting are required. The hydroplane idea promises much in this respect. With a better understanding and control of the conditions associated with successful and safe descent -- perhaps with improved appliances therefor — the problem of ascent will also be partly solved. If such result can be achieved, these measures of control must be made automatic.

The building of complete aeroplanes to standard designs would be extremely profitable at present prices, which range from $2500 to $5000. Perhaps the most profitable part would be in the building of the motor. The framing and fabric of an ordinary monoplane could easily be constructed at a cost below $300. The propeller may cost $50 more. The expense for wires, ropes, etc., is trifling; and unless special scientific instruments and accessories are required, all of the rest of the value lies in the motor and its accessories. Within reasonable limits, present costs of motors vary about with the horse-power. The amateur designer must therefore be careful to keep down weight and power unless he proposes to spend money quite freely.

THE CASE OF THE DIRIGIBLE

Not very much is being heard of performances of dirigible balloons just at present. They have shown themselves to be lacking in stanchness and effectiveness under reasonable variations of weather. We must have fabrics that are stronger for their weight and more impervious. Envelopes must be so built structurally as to resist deformation at high speeds, without having any greatly increased weight. A cheap way of preparing pure hydrogen gas is to be desired.

Most important of all, the balloon must have a higher speed, to make it truly dirigible. This, with sufficient steering power, will protect it against the destructive accidents that have terminated so many balloon careers. Here again arises the whole question of power in relation to motor weight, though not as formidably as is the case with the aeroplane. The required higher speeds are possible now, at the cost merely of careful structural design, reduced radius of action, and reduced passenger carrying capacity.

Better altitude control will be attained with better fabrics and the use of plane fin surfaces at high speeds. The employment of a vertically-acting propeller as a somewhat wasteful but perhaps finally necessary measure of safety may also be regarded as probable.

GIRAUDON'S WHEEL AEROPLANE

THE ORTHOPTER

The *aviplane, ornithoptère* or *orthopter* is a flying machine with bird-like flapping wings, which has received occasional attention from time to time, as the result of a too blind adherence to Nature's analogies. Every mechanical principle is in favor of the screw as compared with any reciprocating method of propulsion. There have been few actual examples of this type: a model was exhibited at the Grand Central Palace in New York in January of this year.

The mechanism of an orthopter would be relatively complex, and the flapping wings would have to "feather" on their return stroke. The flapping speed would have to be very high or the surface area very great. This last requirement would lead to structural difficulties. Propulsion would not be uniform, unless additional complications were introduced. The machine would be the most difficult of any type to balance. The motion of a bird's wing is extremely complicated in its details — one that it would be as difficult to imitate in a mechanical device as it would be for us to obtain the structural strength of an eagle's wing in fabric and metal, with anything like the same extent of surface and limit of weight. According to Pettigrew, the efficiency of bird and insect flight depends largely upon the elasticity of the wing. Chatley gives the ratio of area to weight as varying from fifty (gnat) to one-half (Australian crane) square feet per pound. The usual ratio in aeroplanes is from one-third to one-half.

About the only advantages perceptible with the orthopter type of machine would be, first, the ability "to start from rest without a preliminary surface glide"; and second, more independence of irregularity in air currents, since the propulsive force is exerted over a greater extent than is that of a screw propeller.

THE HELICOPTER

The *gyroplane* or *helicopter* was the type of flying machine regarded by Lord Kelvin as alone likely to survive. It

BRÉGUET GYROPLANE DURING CONSTRUCTION
(Helicopter type)

lifts itself by screw propellers acting vertically. This form was suggested in 1852. When only a single screw was used, the whole machine rotated about its vertical axis. It was attempted to offset this by the use of vertical fin-planes: but these led to instability in the presence of irregular air currents. One early form had two oppositely-pitched screws driven by a complete steam engine and boiler plant. One of the Cornu helicopters had adjustable inclined planes under the two large vertically propelling screws. The air which slipped past the screws imposed a pressure on the inclined planes which was utilized to produce horizontal movement in any desired direction — if the wind was not too adverse. A gasoline engine was carried in a sort of well between the screws.

The helicopter may be regarded as the limiting type of aeroplane, the sail area being reduced nearly to zero; the wings becoming mere fins, the smaller the better. It therefore requires maximum motor power and is particularly dependent upon the development of an excessively light motor. It is launched and descends under perfect control, without regard to horizontal velocity. It has very little exposed surface and is therefore both easy to steer and independent of wind conditions. By properly arranging the screws it can be amply balanced: but it must have a particularly stout and strong frame.

The development of this machine hinges largely on the propeller. It is not only necessary to develop *power* (which means force multiplied by velocity) but actual propulsive vertical *force:* and this must exceed or at least equal the whole weight of the machine. From ten to forty pounds of lifting force per horse-power have been actually attained: and with motors weighing less than five pounds there is evidently some margin. The propellers are of special design, usually with very large blades. Four are commonly used: one, so to speak, at each "corner" of the machine. The helicopter is absolutely dependent upon its motors. It cannot descend safely if the power fails. If it is to do anything but ascend and descend it must have additional propulsive machinery for producing horizontal movement.

Composite Types

The aeroplane is thus particularly weak as to stability, launching, and descending: but it is economical in power because it uses the air to hold itself up. The dirigible balloon is lacking in power and speed, but can ascend and descend safely, even if only by wasteful methods; and it can carry heavy weights, which are impossible with the structurally fragile aeroplane. The helicopter is wasteful in power, but is stable and sure in ascending and descending, providing only that the motor power does not fail.

Why, then, not combine the types? An aeroplane-dirigible would be open to only one objection: on the ground of stability. The dirigible-helicopter would have as its only disadvantage a certain wastefulness of power, while the aeroplane-helicopter would seem to have no drawback whatever.

All three combinations have been, or are being, tried. An Italian engineer officer has designed a balloon-aeroplane. The balloon is greatly flattened, or lens-shaped, and floats on its side, presenting its edge to the horizon — if inclination be disregarded. With some inclination, the machine acts like an aeroplane and is partially self-sustaining at any reasonable velocity.

The use of a vertically-acting screw on a dirigible combines the features of that type and the helicopter. This arrangement has also been the subject of design (as in Captain Miller's flexible balloon) if not of construction. The combination of helicopter and aeroplane seems especially promising: the vertical propellers being employed for starting and descending, as an emergency safety feature and perhaps for aid in stabilizing. The fact that composite types of flying machine have been suggested is perhaps, however, an indication that the ultimate type has not yet been established.

What is Promised

The flying machine will probably become the vehicle of the explorer. If Stanley had been able to use a small high-powered dirigible in the search for Livingstone, the journey would have been one of hours as compared with months, the food and general comfort of the party would have been equal in quality to those attainable at home, and the expense in money and in human life would have been relatively trifling.

Most readers will remember the fate of Andrée, and the projected polar expeditions of Wellman in 1907 and 1909. Misfortune accompanied both attempts; but one has only to read Peary's story of the dogged tramp over the Green-

WELLMAN'S AMERICA
(From Wellman's *Aerial Age*)

land ice blink to realize that danger and misfortune in no less degree have accompanied other plans of Arctic pioneering. With proper design and the right men, it does not seem unreasonable to expect that a hundred flying machines may soar above Earth's invisible axial points during the next dozen years.*

*The high wind velocities of the southern circumpolar regions may be an insurmountable obstacle in the Antarctic. Yet Mawson expects to take with him a 2-passenger monoplane having a 180-mile radius of action on the expedition proposed for this year.

The report of Count Zeppelin's Spitzbergen expedition of last year has just been made public. This was undertaken to ascertain the adaptability of flying machines for Arctic navigation. Besides speed and radius of action, the conclusive factors include that of freedom from such breakdowns as cannot be made good on the road.

For exploration in other regions, the balloon or the aeroplane is sure to be employed. Rapidity of progress without fatigue or danger will replace the floundering through swamps, shivering with ague, and bickering with hostile natives now associated with tropical and other expeditions. The stereoscopic camera with its scientific adjuncts will permit of almost automatic map-making, more comprehensive and accurate than any now attempted in other than the most settled sections. It is not too much to expect that arrangements will be perfected for conducting complete topographical surveys without more than occasional descents. If extremely high altitudes must be attained — over a mile — the machines will be of special design; but as far as can now be anticipated, there will be no insurmountable difficulties. The virgin peaks of Ruwenzori and the Himalayas may become easily accessible — even to women and children if they desire it. We may obtain direct evidence as to the contested ascent of Mt. McKinley. A report has been current that a Blériot monoplane has been purchased for use in the inspection of construction work for an oil pipe line across the Persian desert; the aeroplane being regarded as "more expeditious and effectual" than an automobile.

The flying machine is the only land vehicle which requires no "permanent way." Trains must have rails, bicycles and automobiles must have good roads. Even the pedestrian gets along better on a path. The ships of the air and the sea demand no improvement of the fluids in which they float. To carry mails, parcels, persons, and even light freight — these applications, if made commercially practicable tomorrow,* would surprise no one; their possibility has already been amply demonstrated. With the dirigible as the transatlantic liner and the aeroplane as the naphtha launch of the air, the whole range of applications is commanded. Hangars and landing stages — the latter perhaps on the roofs of buildings, revolutionizing our domestic architecture — may spring up as rapidly as garages have done. And the aeroplane is potentially (with the exception of the motorcycle) the cheapest of self-propelled vehicles.

Governments have already considered the possibilities of aerial smuggling. Perhaps our custom-house officers will soon have to watch a fence instead of a line: to barricade in two dimensions instead of one. They will need to be provided with United States Revenue aeroplanes. But how are aerial frontiers to be marked? And does a

---

*It seems that tomorrow has come; for an aeroplane is being regularly used (according to a reported interview with Dr. Alexander Graham Bell) for carrying mails in India.

nation own the air above it, or is this, like the high seas, "by natural right, common to all"? Can a flying-machine blockade-runner above the three-mile height claim extraterritoriality?

The flying machine is no longer the delusion of the "crank," because it has developed a great industry. A now antiquated statement put the capitalization of aeroplane manufactories in France at a million dollars, and the development expenditure to date at six millions. There are dozens of builders, in New York City alone, of monoplanes, biplanes, gliders, and models. A permanent exhibition of air craft is just being inaugurated. We have now even an aeronautic "trust," since the million-dollar capitalization of the Maxim, Blériot, Grahame-White firm.

According to the New York *Sun*, over $500,000 has been subscribed for aviation prizes in 1911. The most valuable prizes are for new records in cross-country flights. The Paris *Journal* has offered $70,000 for the best speed in a circling race from Paris to Berlin, Brussels, London, and back to Paris — 1500 miles. Supplementary prizes from other sources have increased the total stake in this race to $100,000. A purse of $50,000 is offered by the London *Daily Mail* for the "Circuit of Britain" race, from London up the east coast to Edinburgh, across to Glasgow, and home by way of the west coast, Exeter, and the Isle of Wight; a thousand miles, to be completed in two weeks, beginning July 22, with descents only at predetermined points. This contest will be open (at an entrance fee of

$500) to any licensee of the International Federation. A German circuit, from Berlin to Bremen, Magdeburg, Düsseldorf, Aix-la-Chapelle, Dresden, and back to the starting point, is proposed by the *Zeitung am Mittag* of Berlin, a prize of $25,000 having been offered. In this country, a comparatively small prize has been established for a run from San Francisco to New York, *via* Chicago. Besides a meet at Bridgeport, May 18-20, together with those to be held by several of the colleges and the ones at Bennings and Chicago, there will be, it is still hoped, a national tournament at Belmont Park at the end of the same month. Here probably a dozen aviators will contest in qualification for the international meet in England, to which three American representatives should be sent as competitors for the championship trophy now held by Mr. Grahame-White. It is anticipated that the chances in the international races favor the French aviators, some of whom — in particular, Leblanc — have been making sensational records at Pau. Flights between aviation fields in different cities are the leading feature in the American program for the year. A trip is proposed from Washington to Belmont Park, *via* Atlantic City, the New Jersey coast, and lower New York bay. The distance is 250 miles and the time will probably be less than that of the best passenger trains between Washington and New York. If held, this race will probably take place late in May. It is wisely concluded that the advancement of aviation depends upon cross-country runs under good control and at reasonable speeds and heights rather than upon exhibition flights in enclosures. It is to be hoped that commercial interests will not be sufficiently powerful to hinder this development.

We shall of course have the usual international championship balloon race, preceded by elimination contests. From present indications Omaha is likely to be chosen as the point of departure.

The need for scientific study of aerial problems is recognized. The sum of $350,000 has been offered the University of Paris to found an aeronautic institute. In Germany, the university at Göttingen has for years maintained an aerodynamic laboratory. Lord Rayleigh, in England, is at the head of a committee of ten eminent scientists and engineers which has, under the authority of Parliament, prepared a program of necessary theoretical and experimental investigations in aerostatics and aerodynamics. Our American colleges have organized student aviation societies and in some of them systematic instruction is given in the principles underlying the art. A permanent aeronautic laboratory, to be located at Washington, D.C., is being promoted.

Aviation as a sport is under the control of the International Aeronautic Federation, having its headquarters at Paris. Bodies like the Royal Aero Club of England and the Aero Club of America are subsidiaries to the Federation. In addition, we have in this country other clubs, like the Aeronautic Society, the United States Aeronautical Reserve, etc. The National Council of the Aero Clubs of

America is a sort of supreme court for all of these, having control of meets and contests; but it has no affiliation with the International body, which is represented here by the Aero Club of America. The Canadian Auto and Aero Club supervises aviation in the Dominion

Aviation has developed new legal problems: problems of liability for accidents to others; the matter of supervision of airship operators. Bills to license and regulate air craft have been introduced in at least two state legislatures.

Schools for instruction in flying as an art or sport are being promoted. It is understood that the Wright firm is prepared to organize classes of about a dozen men, supplying an aeroplane for their instruction. Each man pays a small fee, which is remitted should he afterward purchase a machine. Mr. Grahame-White, at Pau, in the south of France, conducts a school of aviation, and the arrangements are now being duplicated in England. Instruction is given on Blériot monoplanes and Farman biplanes, at a cost of a hundred guineas for either. The pupil is coached until he can make a three-mile flight; meanwhile, he is held partially responsible for damage and is required to take out a "third-party" insurance policy.

There is no lack of aeronautic literature. Major Squier's paper in the *Transactions* of the American Society of Mechanical Engineers, 1908, gave an eighteen-page list of books and magazine articles of fair completeness up to its date; Professor Chatley's book, *Aeroplanes*, 1911, discusses some recent publications; the Brooklyn Public Library in New York issued in 1910 (misdated 1909) a manual of fourteen pages critically referring to the then available literature, and itself containing a list of some dozen bibliographies.

AERIAL WARFARE

THE use of air craft as military auxiliaries is not new. As early as 1812 the Russians, before retreating from Moscow, attempted to drop bombs from balloons: an attempt carried to success by Austrian engineers in 1849. Both

(Photo by Paul Thompson)

contestants in our own War of Secession employed captive and drifting balloons. President Lincoln organized a regular aeronautic auxiliary staff in which one Lowe held the official rank of chief aeronaut. This same gentleman (who had accomplished a reconnaissance of 350 miles in eight hours in a 25,000 cubic foot drifting balloon) was subjected to adverse criticism on account of a weakness for making ascents while wearing the formal "Prince Albert" coat and silk hat! A portable gas-generating plant was employed by the Union army. We are told that General Stoneman, in 1862, directed artillery fire from a balloon, which was repeatedly fired at by the enemy, but not once hit. The Confederates were less amply equipped. Their balloon was a patchwork of silk skirts contributed (one doubts not, with patriotic alacrity) by the daughters of the Confederacy.

It is not forgotten that communication between besieged Paris and the external world was kept up for some months during 1870–71 by balloons exclusively. Mail was carried on a truly commercial scale: pet animals and — the anticlimax is unintended — 164 persons, including M. Gambetta, escaped in some sixty-five flights. Balloons were frequently employed in the Franco-Prussian contest; and they were seldom put *hors de combat* by the enemy.

During our war with Spain, aerial craft were employed in at least one instance, namely, at San Juan, Porto Rico, for reconnoitering entrenchments. Frequent ascents were made from Ladysmith, during the Boer war. The balloons were often fired at, but never badly damaged. Cronje's army was on one occasion located by the aid of a British scout-balloon. Artillery fire was frequently directed from aerial observations. Both sides employed balloons in the epic conflict between Russia and Japan.

A declaration introduced at the second international peace conference at the Hague proposed to prohibit, for a limited period, the discharge of projectiles or explosives from flying machines of any sort. The United States was the only first-class power which endorsed the declaration. It does not appear likely, therefore, that international law will discountenance the employment of aerial craft in international disputes. The building of airships goes on with increasing eagerness. Last year the Italian chamber appropriated $5,000,000 for the construction and maintenance of flying machines.

A press report dated February 4 stated that a German aeronaut had been spending some weeks at Panama, studying the air currents of the Canal Zone. No flying machine may in Germany approach more closely than within six miles of a fort, unless specially licensed. At the Krupp works in Essen there are being tested two new guns for shooting at aeroplanes and dirigibles. One is mounted on an armored motor truck. The other is a swivel-mounted gun on a flat-topped four-wheeled carriage.

The United States battleship *Connecticut* cost $9,000,000. It displaces 18,000 tons, uses 17,000 horse-power and 1000 men, and makes twenty miles an hour. An aeroplane of unusual size with nearly three times this speed, employing from one to three men with an engine of 100 horse-power, would weigh one ton and might cost $5000. A Dreadnought costs $16,000,000, complete, and may last — it is difficult to say, but few claim more than ten years. It depreciates, perhaps, at the rate of $2,000,000 a year. Aeroplanes built to standard designs in large quantities would cost certainly not over $1000 each. The ratio of cost is 16,000 to 1. Would the largest Dreadnought, exposed unaided to the attack of 16,000 flying machines, be in an entirely enviable situation?

An aeroplane is a fragile and costly thing to hazard at one blow: but not more fragile or costly than a Whitehead torpedo. The aeroplane soldier takes tremendous risks; but perhaps not greater risks than those taken by the crew of a submarine. There is never any lack of daring men when daring is the thing needed.

All experience goes to show that an object in the air is hard to hit. The flying machine is safer from attack where it works than it is on the ground. The aim necessary to impart a crippling blow to an aeroplane must be one of unprecedented accuracy. The dirigible balloon gives a larger mark, but could not be immediately crippled by almost any projectile. It could take a good pounding and still get away. Interesting speculations might be made as to the outcome of an aerial battle between the two types of craft. The aeroplane might have a sharp cutting beak with which to ram its more cumbersome adversary, but this would involve some risk to its own stability: and the balloon could easily escape by a quick ascent. It has been suggested that each dirigible would need an aeroplane escort force for its defense against ramming. Any collision between two opposing heavier-than-air machines could not, it would seem, be other than disastrous: but perhaps the dirigible could rescue the wrecks. Possibly gas-inflated life buoys might be attached to the individual combatants. In the French manœuvers, a small aeroplane circled the dirigible with ease, flying not only around it, but in vertical circles over and under it.

7.5 CENTIMETER GERMAN AUTOMATIC GUN FOR ATTACKING AIRSHIPS
(From Brewer's *Art of Aviation*)

The French war office has exploited both types of machine. In Germany, the dirigible has until recently received nearly all the attention of strategists: but the results of a recent aerial war game have apparently suggested a change in policy, and the Germans are now, without neglecting the balloon, actively developing its heavier-than-air competitor. England seems to be muddled as to its aerial policy, while the United States has been waiting and for the most part doing nothing. Now, however, the mobilizations in Texas have been associated with a considerable amount of aeroplane enthusiasm. A half-dozen machines, it is expected, will soon be housed in the aerodrome at San Antonio. Experiments are anticipated in the carrying of light ammunition and emergency supplies, and one of the promised manœuvers is to be the locating of concealed bodies of troops by air scouts. Thirty army officers are to be detailed for aeroplane service this year; five training schools are to be established.

If flying machines are relatively unsusceptible to attack, there is also some question as to their effectiveness *in* attack. Rifles have been discharged from moving balloons with some degree of accuracy in aim; but long-range marksmanship with any but hand weapons involves the mastery of several difficult factors additional to those present in gunnery at sea. The recoil of guns might endanger stability; and it is difficult to estimate the possible effects of a powerful concussion, with its resulting surges of air, in the immediate vicinity of a delicately balanced aerial vessel.

But aside from purely combative functions, air craft may be superlatively useful as messengers. To send despatches rapidly and without interference, or to carry a general 100 miles in as many minutes — these accomplishments would render impossible the romance of a "Sheridan's Ride," but might have a romance of their own. With the new sense added to human equipment by wireless communication, the results of observations may be signaled to friends over miles of distance without intervening permanent connections of however fragile a nature.

Flying machines would seem to be the safest of scouts. They could pass over the enemy's country with as little direct danger — perhaps as unobserved — as a spy in disguise; yet their occupants would scarcely be subjected to the penalty accompanying discovery of a spy. They could easily study the movements of an opposing armed force: a study now frequently associated with great loss of life and hampering of effective handling of troops. They could watch for hostile fleets with relatively high effectiveness (under usual conditions), commanding distant approaches to a long coast line at slight cost. From their elevated position, they could most readily detect hostile submarines threatening their own naval fleet. Maximum effective reconnaissance in minimum time would be their chief characteristic: in fact, the high speeds might actually constitute an objection, if they interfered with thorough observation. But if air craft had been available at Santiago in 1898, Lieutenant Blue's expedition would have been unnecessary, and there would have been for no moment any doubt that Admiral Cervera's fleet was actually bottled up behind the Morro. No besieged fortress need any longer be deprived of communication with — or even some medical or other supplies from — its friends. Suppose that Napoleon had been provided with a flying machine at Elba — or even at St. Helena!

The applications to rapid surveying of unknown ground that have been suggested as possible in civil life would be equally possible in time of war. Even if the scene of conflict were in an unmapped portion of the enemy's territory, the map could be quickly made, the location of temporary defenses and entrenchments ascertained, and the advantage of superior knowledge of the ground completely overcome prior to an engagement. The searchlight and the compass for true navigation on long flights over unknown country would be the indispensable aids in such applications.

During the current mobilization of the United States Army at Texas, a despatch was carried 21 miles on a map-and-compass flight, the round trip occupying less than two hours, and being made without incident. The machine flew at a height of 1500 feet and was sighted several miles off.

A dirigible balloon, it has been suggested, is comparatively safe while moving in the air, but is subjected to severe strains when anchored to the ground, if exposed. It must have either safe harbors of refuge or actual shelter buildings — dry docks, so to speak. In an enemy's country a ravine or even a deep railway cut might answer in an emergency: but the greatest reliance would have to be placed on quick return trips from a suitable base. The balloon would be, perhaps, a more effective weapon in defense than in attack. Major Squier regards a flying height of one mile as giving reasonable security against hostile projectiles in the daytime. A lower elevation

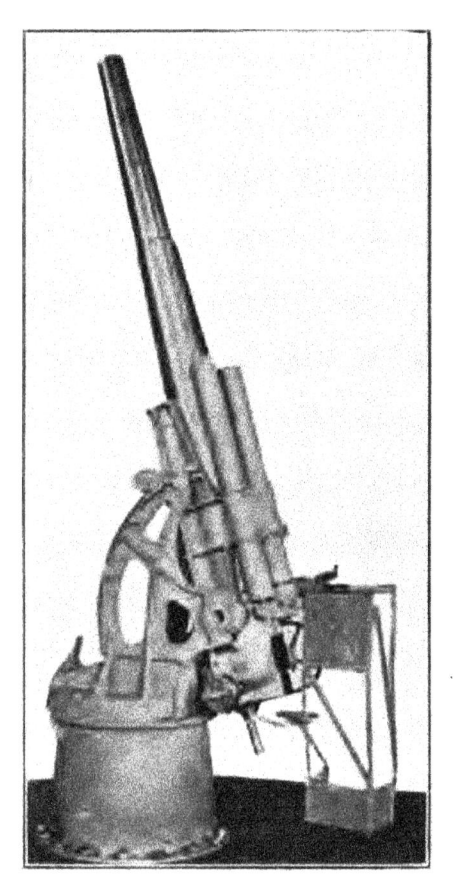

GERMAN GUN FOR SHOOTING AT
AEROPLANES
(From Brewer's *Art of Aviation*)

would be sufficient at night. Given a suitable telephotographic apparatus, all necessary observations could easily be made from this altitude. Even in the enemy's territory, descent to the earth might be possible at night under reasonably favorable conditions. Two sizes of balloon would seem to be indicated: the scouting work described would be done by a small machine having the greatest possible radius of action. Frontiers would be no barrier to it. Sent from England in the night it could hover over a Kiel canal or an island of Heligoland at sunrise, there to observe in most leisurely fashion an enemy's mobilizations.

At the London meeting of the Institute of Naval Architects, in April, 1911, the opinion was expressed that the only effective way of meeting attack from a flying machine at sea would be by a counter-attack from the same type of craft. The ship designers concluded that the aeroplane would no more limit the sizes of battleships than the torpedo has limited them.

For the more serious work of fighting, larger balloons would be needed, with net carrying capacities perhaps upward from one ton. Such a machine could launch explosives and combustibles against the enemy's forts, dry docks, arsenals, magazines, and battleships. It could easily and completely destroy his railroads and bridges; perhaps even his capital itself, including the buildings housing his chief executive and war office staff. Nothing — it would seem — could effectually combat it save air craft of its own kind. The battles of the future may be battles of the air.

There are of course difficulties in the way of dropping missiles of any great size from flying machines. Curtiss and others have shown that accuracy of aim is possible. Eight-pound shrapnel shells have been dropped from an

SANTOS-DUMONT CIRCLING THE EIFFEL TOWER
(From Walker's *Aerial Navigation*)

aeroplane with measurably good effect, without upsetting the vessel; but at best the sudden liberation of a considerable weight will introduce stabilizing and controlling difficulties. The passengers who made junketing trips about Paris on the *Clement-Bayard* complained that they were not allowed to throw even a chicken-bone overboard! But it does not seem too much to expect that these purely mechanical difficulties will be overcome by purely mechanical remedies. An automatic venting of a gas ballonet of just sufficient size to compensate for the weight of the dropped shell would answer in a balloon: a similar automatic change in propeller speed and angle of planes would suffice with the aeroplane. There is no doubt but that air craft may be made efficient agents of destruction on a colossal scale.

A Swedish engineer officer has invented an aerial torpedo, automatically propelled and balanced like an ordinary submarine torpedo. It is stated to have an effective radius of three miles while carrying two and one-half pounds of explosive at the speed of a bullet. One can see no reason why such torpedoes of the largest size are not entirely practicable: though much lower speeds than that stated should be sufficient.

According to press reports, the Krupps have developed a non-recoiling torpedo, having a range exceeding 5000 yards. The percussion device is locked at the start, to prevent premature explosion: unlocking occurs only after a certain velocity has been attained.

Major Squier apparently contends that the prohibition of offensive aerial operations is unfair, unless with it there goes the reciprocal provision that a war balloon shall not be fired at from below. Again, there seems to be no good reason why aerial mines dropped from above should be forbidden, while submarine mines — the most dangerous naval weapons — are allowed. Modern strategy aims to capture rather than to destroy: the manœuvering of the enemy into untenable situations by the rapid mobilization of troops being the end of present-day highly organized staffs. Whether the dirigible (certainly not the aeroplane) will ever become an effective vehicle for transport of large bodies of troops cannot yet be foreseen.

Differences in national temper and tradition, and the conflict of commercial enterprise, perhaps the very recentness of the growth of a spirit of national unity on the one hand, are rapidly bringing the two foremost powers of Europe into keen competition: a competition which is resulting in a bloodless revolution in England, necessitated by the financial requirements of its naval program. Germany, by its strategic geographical position, its dominating military organization, and the enforced frugality, resourcefulness, and efficiency of its people, possesses what must be regarded as the most invincible army in the world. Its avowed purpose is an equally invincible navy. Whether the Gibraltar-Power can keep its ascendancy may well be doubted. The one doubtful — and at the same time perhaps hopeful — factor lies in the possibilities of aerial navigation.

If one battleship, in terms of dollars, represents 16,000 airships, and if one or a dozen of the latter can destroy the

LATHAM, FARMAN, AND PAULHAN

former — a feat not perhaps beyond the bounds of possibility — if the fortress that represents the skill and labor of generations may be razed by twoscore men operating from aloft, then the nations may beat their spears into pruning-hooks and their swords into plowshares: then the battle ceases to hinge on the power of the purse. Let war be made so costly that nations can no more afford it than sane men can wrestle on the brink of a precipice. Let armed international strife be viewed as it really is — senseless as the now dying duello. Let the navy that represents the wealth, the best engineering, the highest courage and skill, of our age, be powerless at the attack of a swarm of trifling gnats like Gulliver bound by Lilliputians — what happens then? It is a *reductio ad absurdum*. Destructive war becomes so superlatively destructive as to destroy itself.

There is only one other way. Let the two rival Powers on whom the peace of the world depends settle their difficulties — surely the earth must be big enough for both! — and then as one would gently but firmly take away from a small boy his too destructive toy rifle, spike the guns and scuttle the ships, their own and all the rest, leaving to some unambitious and neutral power the prosaic task of policing the world. Here is a work for red blood and national self-consciousness. If war were ever needed for man's best development, other things will answer now. The torn bodies and desolated homes of millions of men have paid the price demanded. No imaged hell can surpass the unnamed horrors that our fathers braved.

"Enforced disarmament!" Why not? Force (and public opinion) have abolished private duels. Why not national duels as well? Civilization's control of savagery always begins with compulsion. For a generation, no first-class power has had home experience in a serious armed conflict. We should not willingly contemplate such experience now. We have too much to do in the world to fight.

The writer has felt some hesitancy in letting these words stand as the conclusion of a book on flying machines: but as with the old Roman who terminated every oration with a defiance of Carthage, the conviction prevails that no other question of the day is of comparable importance; and on a matter of overwhelming consequence like this no word can ever be out of place. The five chief powers spent for war purposes (officially, as Professor Johnson puts it, for the "preservation of peace") about $1,000,000,000 in the year 1908. In the worst period of the Napoleonic operations the French military and naval budget was less than $100,000,000 annually. Great Britain, on the present peace footing, is spending for armament more rapidly than from 1793 to 1815. The gigantic "War of the Spanish Succession" (which changed the map of Europe) cost England less than a present year's military expenditure. Since the types for these pages have been set, the promise of international peace has been distinctly strengthened. President Taft has suggested that as, first, questions of individual privilege, and, finally, even those of individual honor, have been by common consent submitted to adjudication, so also may those so-called "issues involving national honor" be disposed of without dishonor by international arbitration. Sir Edward Grey, who does not hesitate to say that increase of armaments may end in the destruction of civilization unless stopped by revolt of the masses against the increasing burdens of taxation, has electrified Europe by his reception of the Taft pronouncement. England and the United States rule one-third the inhabitants of the earth. It is true that a defensive alliance might be more advantageous to the former and disagreeably entangling to the latter; but a binding treaty of arbitration between these powers would nevertheless be a worthy climax to our present era. And if it led to alliance against a third nation which had refused to arbitrate (led — as Sir Edward Grey suggests — by the logic of events and not by subterranean device) would not such be the fitting and conclusive outcome?

The Taft-Grey program — one would wish to call it that — has had all reputable endorsement; in England, no factional opposition may be expected. Our own jingoes are strangely silent. Mr. Dillon's fear that compulsory disarmament would militate against the weaker nations is offset by the hearty adherence of Denmark. A resolution in favor of the establishment of an international police force has passed the House of Commons by a heavy majority. It looks now as if we might hope before long to re-date our centuries. We have had Olympiads and Years of Rome, B.C. and A.D. Perhaps next the dream of thoughtful men may find its realization in the new (and, we may hope, English) prefix, Y.P.— Year of Peace.

Books on Aeronautics

FLYING MACHINES TO-DAY. By WILLIAM D. ENNIS, M. E., Professor of Mechanical Engineering, Polytechnic Institute, Brooklyn.
12mo., cloth, 218 pp., 123 illustrations..........................$1.50 net

CONTENTS: THE DELIGHTS AND DANGERS OF FLYING—Dangers of Aviation—What It is Like to Fly. SOARING FLIGHT BY MAN—What Holds it Up. Lifting Power. Why so Many Sails. Steering. TURNING CORNERS—What Happens When Making a Turn. Lateral Stability. Wing Warping. Automatic Control. The Gyroscope. Wind Gusts. AIR AND THE WIND—Sailing Balloons. Field and Speed. GAS AND BALLAST—Buoyancy in Air. Ascending and Descending. The Ballonet. The Equilibrator. DIRIGIBLE BALLOONS AND OTHER KINDS—Shapes. Dimensions. Fabrics. Framing. Keeping the Keel Horizontal. Stability. Rudders and Planes. Arrangement and Accessories. Amateur Dirigibles. Fort Omaha Plant. Balloon Progress. QUESTION OF POWER—Resistance of Aeroplanes. Resistance of Dirigibles. Independent Speed and Time-table. Cost of Speed. Propellor. GETTING UP AND DOWN ; MODELS AND GLIDERS ; AEROPLANE DETAILS — Launching. Descending. Gliders. Models. Balancing. Weights. Miscellaneous. Things to Look After. SOME AEROPLANES—SOME ACCOMPLISHMENTS. THE POSSIBILITIES IN AVIATION—Case of the Dirigible. The Orthopter. The Helicopter. Composite Types. What Is Promised. AERIAL WARFARE.

AERIAL FLIGHT. Vol. I. Aerodynamics. By F. W. LANCHESTER.
8vo., cloth, 438 pp., 162 illustrations..........................$6.00 net

CONTENTS: Fluid Resistance and Its Associated Phenomena. Viscosity and Skin Friction. The Hydrodynamics of Analytical Theory. Wing Form and Motion in the Peritery. The Aeroplane. The Normal Plane. The Inclined Aeroplane. The Economics of Flight. The Aerofoil. On Propulsion, the Screw Propeller, and the Power Expended in Flight. Experimental Aerodynamics. Glossary. Appendices.

Vol. II. Aerodonetics. By F. W. LANCHESTER.
8vo., cloth, 433 pp., 208 illustrations..........................$6.00 net

CONTENTS: Free Flight. General Principles and Phenomena. The Phugoid Theory—The Equations of the Flight Path. The Phugoid 1852-1872. Dirigible Balloons from 1883-1897 ; 1898-1906. Flying Machine Theory—The Flight Path Plotted. Elementary Deductions from the Phugoid Theory. Stability of the Flight Path as Affected by Resistance and Moment of Inertia. Experimental Evidence and Verification of the Phugoid Theory. Lateral and Directional Stability. Review of Chapters I to VII, and General Conclusions. Soaring. Experimental. Aerodonetics.

AERIAL NAVIGATION. A practical handbook on the construction of dirigible balloons, aerostats, aeroplanes and aeromotors, by FREDERICK WALKER. 12mo., cloth, 151 pp., 100 illustrations..$3.00 net

CONTENTS: Laws of Flight. Aerostatics. Aerostats. Aerodynamics. Screw Propulsion. Paddles and Aeroplanes. Motive Power. Structure of Air-Ships and Materials. Air-Ships. Appendix.

AEROPLANE PATENTS. By ROBERT M. NEILSON. 8vo., cloth, 101 pp., 77 illustrations..........................$2.00 net

CONTENTS: Advice to Inventors. Review of British Patents. British Patents and Applications for Patents from 1860 to 1910, Arranged in Order of Application. British Patentees, Arranged Alphabetically. United States Patents from 1896 to 1909, Arranged in Order of Issue. United States Patentees, Arranged Alphabetically.

(OVER)

THE PRINCIPLES OF AEROPLANE CONSTRUCTION. By RANKIN KENNEDY, C. E. 8vo., cloth, 145 pp., 51 diagrams..........**$1.50 net**

CONTENTS: Elementary Mechanics and Physics. Principles of Inclined Planes. Air and Its Properties. Principles of the Aeroplane. The Curves of the Aeroplane. Centers of Gravity: Balancing; Steering. The Propeller. The Helicoptéré. The Wing Propeller. The Engine. The Future of the Aeroplane.

HOW TO DESIGN AN AEROPLANE. By HERBERT CHATLEY. 16mo., boards, 109 pp., illustrated (Van Nostrand's Science Series)....**50 cents**

CONTENTS: The Aeroplane. Air Pressure. Weight. Propellers and Motors. Balancing. Construction. Difficulties. Future Developments. Cost. Other Flying-Machines (Gyroplane and Orinthoptere).

HOW TO BUILD AN AEROPLANE. By ROBERT PETIT. Translated from the French by T. O'B. Hubbard and J. H. Ledeboer. 8vo., cloth, 131 pp., 93 illustrations..**$1.50 net**

CONTENTS: General Principles of Aeroplane Design. Theory and Calculation. Resistance, Lift, Power, Calculations for the Design of an Aeroplane, Application of Power, Design of Propeller, Arrangements of Surfaces, Stability, Center of Gravity, etc. Materials. Construction of Propellers. Arrangements for Starting and Landing. Controls. Placing Motor. The Planes. Curvatures. Motors.

AIRSHIPS, PAST AND PRESENT. Together with chapters on the use of balloons in connection with meteorology, photography, and the carrier pigeon. By A. HILDEBRANDT, Captain and Instructor in the Prussian Balloon Corps. Translated by W. H. Story. 8vo., cloth, 361 pp., 222 illustrations..**$3.50 net**

CONTENTS: Early History of the Art. Invention of the Air Balloon. Montgolfieres, Charlieres, and Rozieres. Theory of the Balloon. Development of the Dirigible Balloon. History of the Dirigible Balloon, 1852-1872. Dirigible Balloons from 1883-1897; 1898-1906. Flying Machines. Kites. Parachutes. Development of Military Ballooning. Ballooning in Franco-Prussian War. Modern Organization of Military Ballooning in France, Germany, England and Russia. Military Ballooning in Other Countries. Balloon Construction and the Preparation of the Gas. Instruments. Ballooning as a Sport. Scientific Ballooning. Balloon Photography. Photographic Outfit for Balloon Work. Interpretation of Photographs. Hectography by Means of Kites and Rockets. Carrier Pigeons for Balloons. Balloon Law.

D. VAN NOSTRAND CO., Publishers
23 MURRAY and 27 WARREN STREETS, NEW YORK

RETURN TO the circulation desk of any
University of California Library

or to the

NORTHERN REGIONAL LIBRARY FACILITY
Bldg. 400, Richmond Field Station
University of California
Richmond, CA 94804-4698

ALL BOOKS MAY BE RECALLED AFTER 7 DAYS
- 2-month loans may be renewed by calling (510) 642-6753
- 1-year loans may be recharged by bringing books to NRLF
- Renewals and recharges may be made 4 days prior to due date

DUE AS STAMPED BELOW

DD20 15M 4-02

CPSIA information can be obtained
at www.ICGtesting.com
Printed in the USA
BVHW081858191118
533535BV00016B/178/P